ZAPPED

Why Your Cell Phone *Shouldn't* Be Your
Alarm Clock *and* 1,268 Ways to *Outsmart*
the Hazards of Electronic Pollution

· · · · · · · · · ·

ANN LOUISE GITTLEMAN

HarperOne
An Imprint of HarperCollinsPublishers

HarperOne

ZAPPED: *Why Your Cell Phone Shouldn't Be Your Alarm Clock and 1,268 Ways to Outsmart the Hazards of Electronic Pollution*. Copyright © 2010 by Ann Louise Gittleman. All rights reserved. Printed in the United States of America. No part of this book may be used or reproduced in any manner whatsoever without written permission except in the case of brief quotations embodied in critical articles and reviews. For information, address HarperCollins Publishers, 10 East 53rd Street, New York, NY 10022.

HarperCollins books may be purchased for educational, business, or sales promotional use. For information please write: Special Markets Department, HarperCollins Publishers, 10 East 53rd Street, New York, NY 10022.

HarperCollins website: http://www.harpercollins.com

HarperCollins®, 📖 ®, and HarperOne™ are trademarks of HarperCollins Publishers

FIRST HARPERCOLLINS PAPERBACK EDITION PUBLISHED IN 2011

Designed by Ralph Fowler
Illustration work by Kris Tobiassen

Library of Congress Cataloging-in-Publication Data
  Gittleman, Ann Louise.
  Zapped : why your cell phone shouldn't be your alarm clock and 1,268 ways
to outsmart the hazards of electronic pollution / by Ann Louise Gittleman.
      p. cm.
  ISBN 978–0–06–186428–5
  1. Electromagnetism—Health aspects.   2. Electromagnetism—Toxicology.
I. Title.
  RA569.3.G58 2010
  612'.01442—dc22   2010014749

11 12 13 14 15   RRD (H)   10 9 8 7 6 5 4 3 2 1

Praise for *Zapped*

"At our clinic we have been studying electromagnetic sensitivity for over twenty years. This problem is affecting much of our population and we have noticed a significant increase in patients who have chemical and electrical sensitivity. Often the two entities go together, but at times the EMF can be isolated. . . . Ann Louise has clearly laid out the problems and some of the protective modalities that can be used for protection against EMF sensitivity."

—William J. Rea, M.D., FACS, FAAEM

"The book caught my attention immediately because I've been reading about, studying, and writing about the effects of electropollution for several years. This book just adds further confirmation that we live in an 'electropolluted' world, and we need to protect ourselves the best we can. . . . You can learn many, many more simple and more in-depth tips on how to protect yourself by reading *Zapped*."

—Susan M. Lark, M.D., author of *Dr. Susan Lark's Hormone Revolution* and editor of *Women's Wellness Today*

"One of the best ways to lower your exposure to electronic pollution, and improve your health, is to replenish the nutrients they destroy. In *Zapped*, Gittleman lists the products and websites that contain solutions to electrical contaminants. She also goes beyond what anyone has previously discussed: little-known nutrient solutions."

—Nan Kathryn Fuchs, Ph.D., editor-in-chief of *Women's Health Letter*

"Gittleman excels at making critical health issues accessible to people who want to actively take steps to support health and prevent illness. *Zapped* offers a solid education in the health effects of electromagnetic fields, offering guidance on how we can live safely and minimize known DNA damage in an increasingly wireless world. Be forewarned and be forearmed!"

—Camilla Rees, MBA, founder of ElectromagneticHealth.org and Campaign for Radiation Free Schools

"The information contained in this book is extremely interesting, easy to understand, and very pertinent to our health and well-being. Gittleman doesn't suggest that we leave all modern technology behind; instead, she shows us how we can live more safely within the matrix. *Zapped* is a must-read for everyone living in this modern age."

—*Price-Pottenger Journal of Health and Healing*

Also by Ann Louise Gittleman

This book is dedicated to all the canaries among us

who can learn to fly like eagles.

"Facts don't cease to exist just
because they are ignored."

—*Aldous Huxley*

CONTENTS

THE BODY ELECTRIC

Quick. Look around you. What is the most fascinating high-tech thing you use?

Is it your iPhone or BlackBerry? Or that GPS mounted on your windshield? Perhaps it's your Nintendo Wii video game system or your home theater with its monster-size plasma TV and surround-sound speaker system.

With all the latest technological innovations, you may be hard-pressed to come up with your personal favorite digital or electronic appliance or wireless gadget. The truth is, nothing man-made even comes close to the wonder of the human body as an exquisitely tuned and sensitive electrical being.

You might be surprised to know that everything your body did today was made possible by electricity. The organic computer, known as your brain, that runs the whole show emits waves that are electrical. All the sensory information (like hunger and pain) it sends and receives is electrical. This includes the neurons that discharge when you move a muscle, the signals that tell your body to heal a wound, and even the beat of your heart.

In fact, the electricity that facilitates the heart is produced when charged increments of energy—known as ions—flow through the heart and cause contractions. You may remember these ions—calcium, potassium, chloride, and sodium (which are especially high in conductivity). They are found on that Periodic Table we learned about in high school science. Well, every one of the elements on the Periodic Table carries an electrical charge, so we are, in a very real sense, electrical beings in an electrical world.

As Yale scientist Harold Saxton Burr—who first used a voltmeter to test the electromagnetic fields in the human body in 1936—once said, "Electricity is the way nature behaves." When an electric current flows through a conductor or wire, a magnetic field is created; thus, the term *electromagnetic fields* (EMFs). Electric and magnetic fields both interact with the body in various ways. When your heart pumps electrically charged blood (the electric current) through your circulatory system (the conductor), the process creates a powerful magnetic field that surrounds your body and can be measured.

The electromagnetism of the body is so crucial to our functioning that even conventional medicine uses it on a daily basis in diagnostic testing. Every single electron, cell, tissue, and organ in your body carries a very specific frequency or range of frequencies—an electronic signature—that can be measured. This principle is utilized in some of the most accurate and life-saving diagnostic procedures from the electrocardiogram for the heart, the electroencephalogram for the brain, and magnetic resonance imaging (MRI) for the body.

MRIs work through the transference of energy via a specific frequency. The energy in this case is made up of radio waves. The atoms in the body or body parts that are being assessed only absorb these energies if their resonant frequency is a match for the frequency of the energy.

So, it is time to think of yourself as much more than a physical being. You are an energy being as well. And you are the most fascinating high-tech "machine" you use on a daily basis!

MODERN MEDICINE MEETS ANCIENT MEDICINE

The decade of the 1980s ushered in a new awareness for many of us regarding the electrical nature of the body. Such awareness is truly essential to understanding how invisible man-made sources of frequencies from power lines, electrical appliances, and electronic devices can impact our physical health. Two pioneering researchers in the field of energetic medicine, Robert O. Becker, M.D., and Richard Gerber, M.D., were largely responsible for bringing this new awareness to those who were ready to receive it. Both authored books dealing with electromagnetism as the basic life force and its role in health and disease. Those books became time-honored classics—Becker's *The Body Electric*, first released in 1985, and Gerber's *Vibrational Medicine*, released three years later. The information in these books laid a foundation for understanding the electric nature of the body and related ancient healing arts, such as acupuncture, which deal with energetic balance.[1]

Doctors of Oriental Medicine teach that the life force energy, which they call *chi* (Indian yogis call it *prana*, Hippocrates called it natural life force, and Christ called it light), flows through the body in a given sequence through long, narrow energy channels known as meridians. There are two polarizing forces, known as yin and yang through which chi manifests itself. Yin and yang are paired opposites, which express the female and male principles, respectively, that exist in all things in varying degrees. Our state of health, according to Traditional Chinese Medicine, is determined and measured by the balance of yin and yang forces in our bodies.

Balance—energetic balance, which expresses itself physically as chemical balance—is the key to health. When the energy balance is disturbed, chi no longer flows freely through the body. It may become stagnant or flow too forcefully through any of the major meridians, each of which is associated with an organ or set of organs. These imbalances are seen as giving rise to (and preceding development of) physical symptoms in the physical body.

Therefore, before physical symptoms manifest—sometimes years before—an energy disturbance exists, which will eventually express itself in bodily illness if not corrected. To prevent this physical expression of imbalance as disease, the Doctor of Oriental Medicine uses acupuncture to redistribute disturbed energies in the body through insertion of needles at key points along the meridians (points that serve as booster amplifiers to bolster the energy flow).

MODERN SCIENCE MEETS ANCIENT MEDICINE

Western medicine has long been skeptical about acupuncture. Such skepticism was largely based on the inability to think of the body in electrical rather than just biochemical terms. Furthermore, there was no proof of any physical structures that corresponded to the meridians through which the invisible chi was said to flow. Little did conventional health practitioners know that the actual physical structures had already been identified. In the 1960s Korean researcher Kim Bong Han discovered a fine duct-like tubule system that corresponded to the meridian paths as charted in Chinese medicine thousands of years earlier. (Gerber made this stunning disclosure in *Vibrational Medicine*.) This correspondence was later verified by tracking the movement of radioactive isotopes through the tubule system with a high-speed computed tomography (CAT) scan. The isotopes were found to travel the course of traditional acupuncture meridians when injected into acupuncture points. Injection into random, non-acupuncture points did not have this effect.

Additional confirmation of acupuncture came in 1988, when William A. Tiller, Ph.D., of Stanford University found that acupuncture points were measurably different than other points on the body. At the exact center of the acupuncture points—and not at other locations on the body—there is a nearly twenty-five-fold decrease in electrical resistance. In other words, acupuncture meridians are more conductive than other points on the body.[2] It's all about energy and electricity . . . and the magnetic force that comes with it—electromagnetism.

MODERN PHYSICS MEETS ANCIENT MEDICINE

The findings of modern physics are now echoing those of Traditional Chinese Medicine. Both suggest that our bodies are essentially a composite of billions or even trillions of frequencies. These frequencies are expressed in the form of cells, organs, and tissues, which are constantly vibrating and communicating with each other—and the external environment.

How so?

The body is so amazingly sensitive that its bioelectric makeup can be affected by the planets and other celestial bodies. According to electromedicine expert James Oschman, Ph.D., "Sunspots and the cycles of the moon cause changes in ionospheric currents and geophysical fields, which in turn influence the fields within us." Oschman believes that it should come as no surprise that geopathic disturbances can affect human physiology when you consider that geomagnetic storms on the sun can be so intense that they damage satellites, power lines, and telephone cables and disrupt radio communications.[3]

In early 2009, the National Academy of Sciences issued a report that estimated that a major solar storm could cause as much as two trillion dollars' worth of just the initial damages to our communications systems, which might require four to ten years for recovery. In fact, a huge solar storm in 1859 knocked out telegraph communication and caused wires to burst into flames. In March 1989, a smaller storm knocked out power to nine million people in Quebec. The most recent report speculated that a more powerful space storm could, within a few hours, affect water distribution, perishable foods and medication, heating and air conditioning, and everything else that relies on electric power.[4]

Clearly our bioelectrical makeup can be affected by the larger magnetic field of the earth and other celestial bodies. Consider that more babies are born during the full moon due to the moon's gravitational pull on amniotic fluid—similar to the way in which the moon impacts the tides of the earth. Cardiologist Stephen Sinatra, M.D., has noted that there is an increase in chest pain and arrhythmias especially around the full moon or intensified solar flares.[5] There's an increased cancer incidence—mostly skin cancer—among airline pilots who fly close to the radioactive emissions from solar flares. Though no studies have conclusively linked their cancers to cosmic radiation, airline pilots are considered "radiation workers" for the purpose of measuring their occupational safety from the X-rays and gamma rays produced by the sun.[6]

What that means is that we are wired to respond to the electromagnetic forces in the universe, from the fields surrounding far-off celestial bodies to the vibes we pick up from each other to the radio waves from the thousands of cell towers that dot the landscape. The earth itself behaves like a giant electrical circuit. Because animals are also wired this way, they respond to changes within the earth's surface (like earthquakes and tsunamis) even before scientific instruments register these events.

Becker, double-nominee for the Nobel Prize in Physics, compares in *The Body Electric* the meridians of acupuncture to electrical transmission lines, noting that they use direct current to transmit signals of injury. In *Cross Currents: The Perils of Electropollution,* he compares electromagnetic resonance in the human body to MRIs, noting that the body's innate resonances could be used to explain health problems and heal them.[7]

Becker's ideas and words were visionary. We now know that communication between cells of the body is facilitated by biophotons, biological light particles, which represent the quantum (smallest unit) of electromagnetic radiation. Biophoton emissions of an organism, it has been found, reflect the health of that organism. The biophotons of healthy people (and animals and plants) are strong and highly organized. People who are sick have weak and chaotic biophoton radiations. Such radiations are signals of dysfunction and imbalance throughout the body, which occur as the oscillatory rate of cells becomes disturbed. The biophotons even produce a visual image that can be captured through Kirlian photography or imaging techniques like bioliminal photography.

Biophotons are also directed through the body's system of meridians to specific organs or tissues that need them. All of our physical functions, as well as thoughts, emotions, and actions, are accompanied by biophoton communication between cells. These subtle energies operate at lightning speed, much faster than chemical reactions or transmissions of nerve impulses.

Researcher Masaru Emoto, Ph.D., who was trained in alternative medicine, published a book of photographs of water crystals he had exposed to negative and positive emotions. Those photographs, taken through a dark field microscope, show the perfect crystals of the water that had been exposed to emotions like love, words like gratitude, or thoughts of Mother Teresa, and the twisted, tortured crystals exposed to hateful words such as "You make me sick, I will kill you" or thoughts of Adolph Hitler. The photos were featured in the 2004 film *What the Bleep Do We Know!?* Similarly, when positive thoughts are introduced into the biofield of an individual, the quantity and quality of photons emitted by cells increase. Just like the water, it is the opposite when the thought or reaction is negative.

THE STRESS FREQUENCY

It's important for us to know that biophoton signals are blocked when we're under the kind of subliminal stress caused by electropollution. The bottom line is that electropollution is continually disturbing—whether you know it or not—your sympathetic nervous system, which elevates your fight-and-flight response that in turn raises cortisol, your stress hormone. Fluctuations in cortisol lead to numerous health disorders ranging from belly fat and thinning skin to even more serious health problems like erratic sleep patterns, accelerated aging, reduced immunity, cardiovascular disease, blood sugar ups and downs, autoimmune disease, and mood disturbances. For most all of us today, unnatural exposure to artificial frequencies is constant, and often unavoidable, due to the rapidly escalating wireless nature of our society within the past decade.

Some radiation is more damaging than others, depending upon its frequency and proximity. Interestingly, there is one major insight that researchers Emoto, Becker, and Gerber share: there are frequencies that heal, and there are frequencies that harm. Unfortunately, we have been engulfed in a growing maze of accumulating harmful frequencies over the past hundred years, primarily in the extremely low frequency (ELF) range (where power line frequencies fall), the radio frequency (RF)/microwave range (where all things wireless live), intermediate frequencies (dirty electricity), and in the highest frequency ranges of ionizing radiation (such as X-rays and gamma rays). Man-made frequencies in these ranges impact us most basically at an energetic level, at the biophoton level, for we are essentially energetic beings. One of the most significant changes since the 1990s has been the advent of digital communications that work on vibration frequencies that our cells can sense and respond to.

Electromagnetic radiation spans a continuum of vibration rates or frequencies from zero vibrations per second (no vibration) to the cosmic radiation at one hundred sixty billion vibrations per second. Another way of describing this is with wavelength. The wavelength is the distance between successive crests of a repeating vibrational. Wavelength is related to frequency, and wavelength goes down as frequency increases.

The relationship between different radiations is shown in a chart called the electromagnetic spectrum. That spectrum is divided into ionizing radiation on the top (which consists of very short, high-energy waves that are powerful enough to damage cellular matter) and nonionizing radiation below X-rays (which consists of longer, lower-energy waves that are still powerful enough to have a significant impact on matter) at the bottom. The dividing line between ionizing radiation and nonionizing radiation falls just above visible light. Each section of the electromagnetic spectrum (see chart) has a specific frequency expressed in vibration or cycles per second which has been given the name Hertz.

Like everything else in our world, our bodies and every organ and tissue they contain have their own distinct frequency. The late Bruce Tainio of Tainio Technology, an independent division of Eastern State University in Cheney, Washington, built the first frequency monitor in the world, and using it, he determined that the average frequency of the human

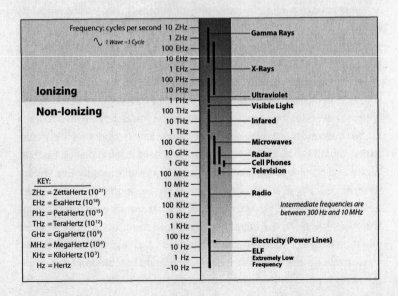

body during the daytime is 62 to 68 Hz. When the frequency drops to 58 Hz, cold and flu symptoms appear; at 55 Hz, disorders like candida take hold; at 52 Hz, Epstein-Barr, and at 42 Hz, cancer.[8]

It seems that in addition to being physical beings, on a more basic level, we are beings of energy whose chemical processes are dependent upon the free flow of that energy through our bodies. Findings in the field of physics have confirmed this, beginning with Einstein's discovery that matter and energy are basically interchangeable.

THE HEALING FREQUENCY

When we take a substance—a drug, an herb, a food, etc.—to treat an illness, and we recover, it is not due to the action of the physical form of the substance but rather to its frequency (the oscillation per second of which it is composed). In other words, it is the frequency that heals, not the physical substance itself.

Nobel Prize winner Günter Blobel, M.D., Ph.D., established that the cells of our bodies send out frequencies (energetic signals) when they need a specific nutrient. These signals are picked up by peptides (which are combinations of two or more amino acids or protein constituents), which act like radio antennas and serve as protein chaperones to escort the needed nutrient to the cell that has sent out the signal for it. They're like your cell's own waiters, taking their orders and bringing them their meal. Nutrition then, like everything else in our world, has an energetic as well as a chemical component.

Many years ago I had the privilege of studying with Hazel Parcells, M.D., Ph.D., who changed the way I perceived and later practiced nutrition. She taught me to think of each cell in the body as an electric battery broadcasting the pulsating rhythm of life. When the energies or vibrations are changed, millions of little batteries (the cells) are growing weaker or stronger. Nothing exists except in its own rate of energy. When you work with the energy, you can change the environment, which in turn changes the expression of health or disease.

THE FINAL FRONTIER

In other words, by fortifying our internal defenses, an impenetrable foundation can be created that will help us to better ward off the latest breed of environmental assaults out there—the unnatural frequencies from all our digital toys and conveniences that we are bombarded with morning, noon, and night.

Let's now take a closer look at electropollution, the electrical and/or magnetic fields generated from all those cell phones, wireless or radio frequency fields, high-tension power lines, wiring in the walls of your home, and common household appliances like your refrigerator and hair dryer. No one can wave a magic wand and immediately remove all the electropollution from our lives. But what this book does hope to accomplish is to present the truth and consequences behind this uncharted territory and offer real-life solutions. But first, you will need to get a better handle on how your world is swirling in a sea of invisible current. The next chapter will do just that.

THE BODY INTERRUPTED

Let's do a little experiment:

Let's take a trip back in time, and then do some fast-forwarding—to give you some idea of the way our technology has grown over the past fifty years and why we're so overexposed to EMFs.

If you're forty or older, close your eyes and think back to your childhood home. If you're under forty, think back to your grandparents' house when you were a kid.

Now take yourself on a mental tour of the house. As you walk from room to room, take a quick visual inventory in your mind of how many electric and electronic appliances and gadgets you see.

If your family was typical, here's what you'll likely come up with:

- **Master bedroom:** A clock-radio or alarm clock, unless it's a windup clock

- **Bedroom #2:** A clock-radio or alarm clock, unless it's a windup clock

- **Bedroom #3:** A clock-radio or alarm clock, unless it's a windup clock

- **Bathrooms:** No appliances, or maybe an electric razor

- **Family room:** A TV, a stereo (maybe), phone

- **Living room:** No appliances

- **Kitchen:** Stove, refrigerator, dishwasher (maybe), blender, can opener, electric knife, toaster, phone

- **Total inventory:** 15

And if you were to take a similar tour of your own home today?

- **Master bedroom:** TV, TiVo, cable box, DVD player, remote control for TV, remote control for TiVo, remote control for cable box, remote control for DVD player, cell phone, cell phone charger, iPod, iPod charger, iPod docking station, remote control for iPod docking station, Bluetooth headset, Bluetooth headset charger, computer, monitor, wireless mouse, wireless keyboard, printer, scanner, digital camera, digital picture frame, air purifier, alarm clock, cordless phone, PDAs

- **Bedroom #2:** TV, TiVo, cable box, DVD player, remote control for TV, remote control for TiVo, remote control for cable box, remote control for DVD player, cell phone, cell phone charger, iPod, iPod charger, iPod docking station, remote control for iPod docking station, Bluetooth headset, Bluetooth headset charger, computer, monitor, wireless mouse, wireless keyboard, printer, scanner, digital camera, digital picture frame, air purifier, alarm clock, cordless phone

- **Bedroom #3:** TV, TiVo, cable box, DVD player, remote control for TV, remote control for TiVo, remote control for cable box, remote control for DVD player, cell phone, cell phone charger, iPod, iPod charger, iPod docking station, remote control for iPod docking station, Bluetooth headset, Bluetooth headset charger, computer, monitor, wireless mouse, wireless keyboard, printer, scanner, digital camera, digital picture frame, air purifier, alarm clock, cordless phone

- **Bedroom #4:** TV, TiVo, cable box, DVD player, remote control for TV, remote control for TiVo, remote control for cable box, remote control for DVD player, cell phone, cell phone charger, iPod, iPod charger, iPod docking station, remote control for iPod docking station, Bluetooth headset, Bluetooth headset charger, computer, monitor, wireless mouse, wireless keyboard, printer, scanner, digital camera, digital picture frame, air purifier, alarm clock, cordless phone

- **Bathroom #1:** TV, iPod docking station, remote control for TV, remote control for iPod docking station, rechargeable electric toothbrush, rechargeable electric razor, curling iron, hair dryer, contact lens cleaner, digital scale/body fat monitor

- **Bathroom #2:** TV, iPod docking station, remote control for TV, remote control for iPod docking station, rechargeable electric toothbrush, rechargeable electric razor, curling iron, hair dryer, contact lens cleaner, digital scale/body fat monitor

- **Family room:** Home theater system (including monster-size flat-panel TV, TiVo, cable box, DVD player, surround-sound speaker system), iPod docking station, remote control for TV, remote control for TiVo, remote control for cable box, remote control for DVD player, remote control for speaker system, remote control for iPod docking station, computer, monitor, wireless mouse, wireless keyboard, printer, scanner, digital camera, digital picture frame, air purifier, digital thermostat, cordless phone, wireless router

- **Living room:** iPod docking station, remote control for iPod docking station, digital picture frame, air purifier, cordless phone, invisible-fence collar on dog, wireless security system

- **Kitchen:** Stove, refrigerator, microwave, dishwasher, blender, toaster oven, food processor, TV, cable box, DVD player, iPod docking station, remote control for TV, remote control for cable box, remote control for DVD player, remote control for iPod docking station, computer, monitor, wireless mouse, wireless keyboard, printer, digital picture frame, air purifier, wall-mounted security system panel, coffeemaker or espresso machine (or both), water filtration system, electronic cat feeder, electronic cat-box cleaner, electronic dog-door/dog-collar system, cordless phone, rechargeable flashlight, rechargeable mini-vac

- **Total inventory:** 192

ZAPPED

So what's really going on here?

We're getting zapped.

If you pay close attention to your activities for just one typical day, you'll quickly realize that a new form of invisible pollution is all around you and, as you'll learn, *within* you, twenty-four hours a day, seven days a week. Granted, you probably don't have all those electronic gizmos in your house, and not all of these modern-day wonders are emitting dangerous radiation. Plus, let's face it, how much time do you spend sitting in front of your electric coffeemaker anyway? How long you're exposed often means more to your health than the actual strength of the electrical, magnetic, or RF field, something that will become very important when we go on our EMF clean-up crusade of every room of your house in a later chapter. But I used this long list to give you some idea of how much life has changed in the digital age—and how many more electromagnetic fields we're exposed to than our grandparents were.

Think about what you did today:

Perhaps you woke to the smell of coffee brewed exactly the way you like it by your electric coffeemaker, which you set on a timer the night before. Maybe you went downstairs, flicked on the fluorescent lights in the kitchen, pulled a frozen breakfast out of the refrigerator, popped it into the microwave, and sipped your first cup of coffee while you waited for it to heat. If you couldn't wait until you got to the office, you pulled your smartphone or cell phone out of its holster and checked your e-mail and then pulled up local traffic and weather reports. And after all of this, you took a hot shower and were thrilled that the new water heater let you take a long, luxurious one.

You might have taken an electric train or subway to work. If you drove, you probably paid no attention to the power lines strung on the ubiquitous wooden poles that are as much a part of the landscape as trees, or the huge transmission poles lumbering across the countryside like sci-fi giants. If you happened to glance over at the car or passenger next to you, you likely saw someone else, like you, on a cell phone, starting the business day before it officially opens. At work, you might walk through an automatic door, take an elevator to your office, flick on the overhead lights, and boot up your computer.

Then, at the end of the day, you reverse it all. Maybe you stop at the supermarket on your way home and buy a few things for dinner, which the checker whisks through a price scanner and tosses in a bag for you. If you're cooking from scratch, you preheat your electric oven, defrost the chicken in the microwave, and put it in to bake. You'll mash the potatoes you boiled on the range with your electric mixer and open the canned green beans with an electric can opener.

Maybe you'll sit in a comfortable automated massage chair before you finish up a report on your laptop computer tonight, or huddle with your eight-year-old while he does his homework and then challenges you to an online game of Scrabble. You might watch a little satellite TV before climbing into bed, where you root around for the remote that controls the firmness or angle of your mattress.

Everything you did, from making coffee to taking a shower to taking the train to buying groceries to going to bed, exposed you in some small or large way to electromagnetic fields, which are invisible force fields that surround all electrical devices. For many people, these invisible energy fields appear to be benign. They have no symptoms—at least, none that they recognize. But others seem acutely attuned to what others can't see, touch, or feel.

FROM ELECTRICITY TO ELECTROPOLLUTION

The widespread use of the light bulb—one of the most life-changing events in the past ten thousand years—was how it all began. In October 1882, Thomas Edison built the first electrical plant that lit just thirteen hundred street lamps and homes in New York City. What followed was an unprecedented avalanche of inventions that harnessed electric power to make Americans more productive and prosperous, as well as safer and healthier, than ever before. In just the first half of the twentieth century, Americans were introduced to everything from conveyor belts, printing presses, electrocardiograms, and X-ray machines to radio, radar, television, and computers.

In the last fifteen years alone, the latest modern electronic wonder—wireless technology—has expanded like a sponge in water. Today, 84 percent of Americans own cell phones, and by 2012, the wireless industry is expected to become a larger sector of the U.S. economy than agriculture and automobiles. About eighty-nine million of us watch TV shows beamed to us by satellite—sports, music, comedy, and drama captured by a metal dish on the roof or outside a high-rise window. And you can't have a cup of coffee at Starbucks without being subject to Wi-Fi, the wireless network that allows you to surf the Internet as you sip your latte.

Yet we may not understand the potential consequences of our latest discoveries any better than our earliest ancestors understood the perils of fire.

For the last ten years in my clinical practice as a researcher, author, and educator, I have been seeing a strange constellation of symptoms in my clients that defy diagnosis and resist even the most tailor-made diet, evidence-based supplementation, state-of-the-art exercise, clinical testing, or even significant lifestyle changes. Consider these very different, but equally puzzling, case histories:

> Fresh out of college, a newlywed moves with her husband to what ought to be the healthiest place in the world—a farm in America's heartland. Yet within six months, this twenty-three-year-old has become so weak, she can barely walk up the stairs. She has developed daily headaches, circulation problems, and hot flashes. She wakes up every morning feeling like she has been "hit by a Mack truck and then run over by a train." Her doctor tells her she has chronic fatigue syndrome.

> While on the job, an emergency room physician suffers from blinding headaches, dizziness, and muscle weakness that make it impossible for her to intubate a patient or even smile for a photo. At one point, her arms and legs turn blue, her vision begins to fade, and her heart

begins to squeeze "as if it was empty." One doctor diagnoses her as mentally ill.

Out of the blue, a high-powered editor, who has been commuting for years to her New York office by train, suddenly begins to feel nauseous during every morning and evening ride. She blames it on stress, but it's becoming so debilitating, she considers quitting the job she loves.

The parents of a young Wall Street trader—who spends most of his working day with a cell phone stuck in each of his ears— are concerned as his health declines over a three-year period, during which he is diagnosed with a host of conditions including autoimmune disorder, parasites, and mercury toxicity.

A recent engineering school graduate working for the Canadian Navy experiences fatigue so severe, he needs to take naps on his lunch hours. As time goes on, he develops chronic respiratory infections, nausea and digestive distress, heart palpitations, and trouble focusing. His diagnosis: stress.

A normally well-behaved thirteen-year-old boy who is doing well in school suddenly develops a behavior problem. Oddly enough, he only acts out at a specific time each day, which mystifies his parents and doctors.

UNRAVELING THE MYSTERY

In the past decade, I have experienced some of these same baffling symptoms for which I too found no relief. In 2005, I was diagnosed with a (thankfully) benign tumor of the parotid, one of the salivary glands located just below the earlobe. Why I got it was a mystery that puzzled even my doctor. It's a very rare tumor, most often caused by radiation exposure.[1] I didn't live near a nuclear plant, I hadn't been exposed to an inordinate number of medical X-rays or other screening tests, and, except for a brief time I spent working as a nutritionist in a hospital, I hadn't even been near a CAT scanner or MRI machine. But, on a hunch, I began my investigations with a theory: what if what these six people and I were suf-

Why Safety Standards Don't Keep Us Safe

There's another factor that may make us vulnerable to EMF effects: public safety standards that are, in a word, obsolete. For one thing, they're based solely on whether an EMF will shock or burn you, the so-called thermal standard which is usually applied to ionizing radiation, like that from microwave ovens and X-rays, which can cause serious burns and are linked with cancer. (Extremely low frequencies, or ELFs, are considered nonionizing radiation, a distinction that simply means they're not strong enough to detach electrons from atoms or molecules.) An increasing number of studies have shown damaging effects far below those levels. In fact, in the earliest studies, rates of leukemia in children living near power lines went up even though the measured field was a thousand times lower than the level deemed safe.

Even the government's own experts have expressed concern about the adequacy of current safety guidelines. In 1999, the Radiofrequency Interagency Working Group—made up of government agencies involved in RF issues—concluded that the RF standard at the time (which remains the same today) "may not adequately protect the public." Among other things, the group cited a lack of study on long-term, low-level chronic exposures as well as the fact that the standards are based only on adult males (as the "average" person), so women, who tend to be smaller on average, and children, who are much smaller and whose

fering from was an environmental condition, one caused by something we're exposed to *every day* but consider harmless?

There are several historical connections that supported my suspicions. Many well-respected historians believe that the Romans were the first society to be destroyed by environmental toxicity. Wealthy Romans painted their walls with lead-based paint. They used the heavy metal for everything, water pipes to toys, statues, cosmetics, coffins, and roofs. But in an article written for *The New England Journal of Medicine,* lead poisoning researcher Jerome Nriagu, Ph.D., D.Sc., an environmental chemist at

maturing brains and bodies are a hotbed of cell activity, may be left extraordinarily vulnerable.[2]

Safety standards for all kinds of EMF exposure got their start in the 1940s and early 1950s as a result of concern for radar operators who were experiencing a wide variety of symptoms, including infertility, internal bleeding, cataracts, headaches, and brain tumors, possibly as a result of their exposure to radar, which sits on the electromagnetic spectrum right between radio frequency and infrared radiation. There are a number of organizations—a variety of industry organizations including utilities and technology manufacturers—that have played a role in developing these so-called safe levels which have been adopted by all of the government agencies charged with overseeing EMF devices, from the Federal Communications Commission (FCC) to the U.S. Food and Drug Administration (FDA). Unfortunately, despite the volume of evidence—even more than I've described here—they have remained resistant to altering safety standards because they do not find the evidence "convincing." As EMF experts and activists Cindy Sage and David Carpenter, M.D., wrote in their report in the August 2009 issue of the journal *Pathophysiology,* "These bodies assume . . . that only conclusive scientific evidence (absolute proof) will be sufficient to warrant change and refuse to take action on the basis of a growing body of evidence which provides early but consequential warning of risks."[3]

After all, the government didn't wait for absolute proof to place dire warnings on cigarette packs in 1966. The link between smoking and diseases such as leukemia, cataracts, pneumonia, aneurysm, and stomach cancer wasn't actually confirmed until 2004. Sometimes watchful waiting just isn't enough.

the University of Michigan, says that it was their consumption of copious amounts of wine that may have given them their heaviest dose.[4]

The Romans flavored their wine by simmering the grape juice in lead pots or lead-lined copper kettles, which not only affected taste but made the wine last longer. Lead has a sweet taste, so it enhanced the sweetness of the wine—which earned the metal the reputation as the sweet poison. The acidic nature of the grapes extracted large amounts of lead from the utensils, and then the Romans quaffed the drink out of lead cups. They may have been taking in as much as 20 mg of lead a day just from wine alone, enough to cause chronic lead poisoning, diminish fertility, and cause mental and emotional impairments.[5]

After more than a year of research, I've come to the conclusion that we, like the ancient Romans, are being exposed to an invisible type of "new" pollution that is making our life "sweeter"—certainly more convenient—but which comes with formidable and unforeseen side effects.

It's called electropollution. It's odorless, colorless, and invisible, and it's probably enveloping you right now. As writer Sara Shannon writes in her 1993 book, *Technology's Curse: Diet for the Atomic Age*, about low-level radiation: "It cannot be seen, felt or heard. It is tasteless and odorless. It is in our food and in the air; it is in our blood and in our bones and can remain in our ashes to go on to contaminate someone else."[6]

Our "sweet poison" is the EMFs produced by our cell phones, PDAs, wireless networks, cell and broadcast towers, power lines, fluorescent lights, even the electrical systems that power our appliances, TVs, computers, and bedside alarm clocks—all those technological devices that make our lives easier. We are affected 24/7 by an unprecedented number of frequencies and wavelengths. By some estimates, we're exposed daily to as much as one hundred million times more electromagnetic radiation than our grandparents were. It flows around us, in us, and interferes with the body's fundamental electric forces of life, including the communication between our cells that tells them how to grow, develop, divide, and even when to die.

Remember those six people I just told you about? They ultimately unraveled the root cause of their mysterious ailments, as I did. They had been zapped.

The newlywed's symptoms were finally traced to a current of electricity that was traveling along the ground and hitchhiked into her home via her own electrical system and water pipes.

The ER doctor, exposed to toxic mold in her home, had developed multiple sensitivities to common everyday chemicals and EMFs.

The editor discovered by accident that if she sat every morning in the train's designated quiet car, away from the cell phones, BlackBerrys, and laptops of her fellow commuters, her symptoms disappeared.

After months of research, the thirteen-year-old's mother found that his sudden behavioral changes coincided with a radar beam sweep of their home from a nearby naval station. The family moved away from the radar beam and the young man's behavioral problems disappeared.

The Wall Street trader was forced to quit his lucrative job to get away from the buzzing hive of techno-gizmos on the trading floor. Today, he may make less money, but he's also symptom-free.

The engineer was able to stay in his job because his employer—the Canadian government—provided him a shielded office to protect him from the radar and other devices that caused his chronic illness.

In my case, years tethered to a computer and cell phone while writing a myriad of books and promoting them on the road had sensitized me to the very tools I depended upon for my career. My parotid tumor turned out to be one of several kinds of tumors linked to cell phone use—and I developed it after several years of traveling constantly and literally living on my cell phone in cars, trains, and planes.[7]

While I haven't given up my cell phone or my computer, you won't find me spending hours on either of them. I've learned to work around—and live well and happily with—modern-day technology.

So, based on my own experience and subsequent research, that is how this book came about. In it, I'm going to share my secrets for peaceful coexistence with everything from your cell phone to the radio frequency towers that make your phone calls possible. I congratulate you for picking up *Zapped*. It means that you're already aware—and perhaps even a little bit concerned—that, despite all the reassurances you've heard, something is simply not right in our world today.

WHY DO I FEEL THIS WAY?

Perhaps you have strange symptoms, like the people I've described here, many of whom you'll meet again later in these chapters. Perhaps you've suspected all along that modern-day maladies like sleeplessness, chronic fatigue syndrome, fibromyalgia, depression, anxiety, and rising rates of cancers and brain tumors, particularly in young people, may have an underlying environmental cause. One eminent researcher has already made that connection. Samuel Milham, M.D., M.P.H., of the Washington State Department of Health, wrote in the journal *Medical Hypotheses* in 2009 that he traced the rise in degenerative disease, cardiovascular disease, and suicide in the United States to the spread of electrical power to urban and rural areas, which was completed around 1956. He compared government disease and mortality statistics before and after electrification and found what is, in essence, the tipping point. When agricultural areas became electrified, rates of these lifestyle diseases started to match that of urban areas, where electricity was introduced in the late 1800s.

"I hypothesize that the 20th century epidemic of the so called diseases of civilization including cardiovascular disease, cancer and diabetes and suicide was caused by electrification not by lifestyle," writes Milham. "A large proportion of these diseases may therefore be preventable."[8]

I agree that they are preventable. If you had enough curiosity to start reading this book in the first place, then I'll bet you have the courage to be proactive, powered by the knowledge to take the necessary steps to be healthier, happier, and to regain your peace of mind. And I can assure you, you'll find most of these steps here—I've taken them myself.

ALLERGIC TO THE DIGITAL WORLD

How have we become allergic to a force that has been with us since time began? Even if you could turn back the hands of time to before 1882 when Edison's electric plant triggered a social, scientific, and industrial revolution, you would still have been exposed to electromagnetic energy.

We, and the universe we live in, produce and operate in a sea of both natural and unnatural electrical and magnetic fields. The earth, for example, pulses at about 10 Hz, like a small engine. Our bodies, as you may remember from chapter 1, are really electromagnetic machines. We simply can't move a muscle or produce a thought without an electrical impulse—and wherever there is electricity, a magnetic field is also produced, which is why we link the two together *electromagnetic.*

As I alluded to in chapter 1, over aeons, our bodies have grown accustomed to the low energy of those natural electromagnetic fields and the wavelengths and frequencies they produce. In fact, they play a positive and important role in all life on earth. Humans have lost most if not all of our awareness of it, but animals still dance to its silent orchestrations. You can see it in their behavior and their ability to foretell earthquakes, hurricanes, and tsunamis—not through any supernatural power but by their acute sensitivity to the earth's electromagnetic hum and electrostatic charges in the air. You can see it in the migratory patterns of birds and animals who seem to be innately directed with some unknown internal antenna.[9]

Many scientists now suspect the secret of their mysterious know-how may be magnetite, a mineral that is a million times more magnetic than iron and is found in the tissue of every living thing: in the eye area of birds, the lines on the bodies of fish, the teeth of sea mollusks, and the abdomens of bees. It links them to the electromagnetic fields of the earth, keeping them plugged in, so to speak, to the earth's energy.[10]

And it's in us too. Small amounts of this magnetic substance are also found in the brain tissue, blood-brain barrier, and the bone above the eyes and sinuses of humans. What effect this internal compass has in us is unknown. We do know that there is a very narrow range of electromagnetic frequencies to which the brain cells of animals and humans respond favorably, and it roughly matches the frequencies produced by the electromagnetic fields produced naturally by our world.[11]

What we are also beginning to understand is that the proliferation of technology, while it has taken us many strides in social and economic progress, may have finally created a toxic load that is too great for some bodies to handle, just as the rapid rise of toxic chemicals such as pesticides, plastics, and heavy metals in the environment has overwhelmed the ability of some bodies to neutralize them. In fact, many experts believe, like I do, that these invisible fields are contributing to making us sick. I am totally convinced that they made *me* sick.

Coming to Terms

The language of electromagnetic fields, much of it from physics, is difficult to navigate, so I've tried to minimize my use of jargon in this book. However, there are certain terms, particularly related to measurement of electric, magnetic, and radio frequency fields, that are impossible to avoid. Here are a few simple definitions that you need to know:

Wave/Wavelength   Electricity is delivered to our homes in alternating current (AC), which means that the electrical charge that flows through the wires periodically reverses direction (cycle), and it's usually shown as an undulating wave, called a sine wave. Wavelength is the measurement of the distance between two peaks of the wave.

Frequency   How many cycles a wave completes in a period of time is known as its frequency. If you live in North America, electrical currents flow to your wires at 60 cycles a second; in Europe, it's 50 cycles a second. This is termed alternating current, or AC for short.

Gauss   This is the measurement unit for magnetic fields. Most prudent scientists today recommend that safe exposure for humans to an AC magnetic field is 1 milliGauss (mG) or less at any single exposure, though other agencies recommend 2 to 3 mG. The earth's magnetic field measures about 0.5 mG.

WHY WE GET ZAPPED

Why are we so vulnerable? The human body, which is 75 percent water, conducts electricity. It's also an effective antenna that picks up energy from the surrounding environment. If you've ever adjusted TV rabbit ears, you know that just the touch of your hand can bring in a better picture. That's because at that moment, the RF waves carrying the image are broadcasting them to you. For that moment, you *are* the antenna.

Hertz   This is a newer term for cycles per second that was awarded to Heinrich Hertz, an early researcher in electromagnetism. The electricity that comes into U.S. homes is 60 Hertz (Hz). Our brainwaves can even be measured in Hertz. For instance, when you are asleep, your brain hums at 1 Hz, or one cycle per second. When you're thinking, whether it is problem solving or being creative—it revs up to as much as 40 Hz.

Extremely Low Frequency Electromagnetic Fields (ELF)   These are the electromagnetic fields in the frequency range of 1 to 30 Hz. Our entire electrical power system and our appliances produce 60 Hz magnetic and electric fields. These fields are a form of nonionizing radiation which won't detach electrons from atoms or molecules. Only ionizing radiation or energy from radioactive substances and cosmic rays are thought to do that: The X-ray you had at the dentist's office and the CAT scan (which uses X-rays) that found your kidney stone both emit ionizing radiation.

Radio Frequency (RF) Field   Another form of nonionizing radiation, these high frequency EMFs are generated by the equipment that transmits wireless signals, such as cell towers, broadcast towers at your local radio or TV stations, and the equipment that receive those signals—your cell or cordless phone. Wireless operates in the microwave band of radio frequency radiation.

In fact, humans are literally walking conversations, cells chattering to one another and interacting with the natural world using electrical charges and chemicals to make the connections. But when you add artificial electromagnetic forces to the mix, we are starting to learn, these quiet conversations suddenly become cacophony, as though a flash mob has arrived, shouting and screaming and with boom boxes blaring so loudly you can't hear yourself think.

And, in essence, that's exactly what happens when you're exposed to the ever-expanding web of electromagnetic forces of varying sizes and strengths that our bodies aren't used to.

Humans have as little protection from this kind of pollution as we do for toxic chemicals, though we do have *some*. For example, the voltage from low-level electrical fields, a form of ELFs, produced by your appliances or overhead power lines, have a limited ability to penetrate the body. Your cell membranes block electrical fields, though not completely.[12]

But your body *will* pick up whatever electrical field you're exposed to, even if it's just the 60 Hz field of your electric shaver or hair dryer; the magnetic field that accompanies it is absorbed entirely. As *New York Times* writer B. Blake Levitt points out in her landmark book *Electromagnetic Fields: A Consumer's Guide to the Issues and How to Protect Ourselves,* this may disrupt your internal electrical field as well as interact with magnetic metals like iron and copper and charged particles in your blood to affect your health in still little-known ways.[13]

CELLS, INTERRUPTED

So what might happen when your own electromagnetic field encounters one outside your body, one that's louder than the ambient levels with which we have evolved? For one thing, it may interfere with the messages your body's cells send and receive—what the late scientist W. Ross Adey, M.D., of the Pettis Memorial Veterans Administration Hospital in Loma Linda, California, referred to as "whispers between cells."[14]

In his experiments, Adey, who chaired the National Council on Radiation Committee on Extremely Low Frequency Electromagnetic Fields, found that both very low frequency fields, such as those produced by our electrical system, as well as the higher radio frequencies utilized by cell phone and broadcast towers may interrupt that cellular chatter or drown out the electrical impulse that carries messages across the cell membranes.[15]

You've probably had the experience of losing a radio station or even picking up another when you drive under a high voltage line or near broadcast towers. Likewise, your cells, exposed to the same electromagnetic forces, may drop their conversation with one another or pick up outside interference that could muddle their messages. It's one thing to miss the chorus of your favorite song, another to have an important thought—*watch out for that car*—get lost in translation because brain cells become confused.

ELECTROMAGNETIC SCIENCE— WHAT HAPPENS IN YOUR BODY

Here's a little bit of the science of what happens when you're over-exposed to EMFs:

Your cells get overwhelmed by messages from inside and outside your body.

One way these artificial fields may disrupt normal electrochemical communication is by increasing the number of what are called receptors on the surface of your cells. Receptors are often described as a keyhole into which the key—a chemical messenger called a neurotransmitter—fits perfectly to open the cell, allowing outside information from your brain or other parts of your body to get inside. These receptors and neurotransmitters help transfer those messages from one cell to the other, like a game of Whisper Down the Lane, but without the garbled transmission.

For example, if you're sick or injured, you want your cells to send out an SOS to your immune system so healing starts right away. And if you're otherwise healthy, that's what happens. The numbers of receptors you have varies widely, and some cells have so many that they're likely to attract more activity than those with fewer receptors.

If overexposure to EMFs boosts receptor numbers, more of your cells become increasingly open to all kinds of messages from your body and from the environment outside. Suddenly, while trying to send its vital 911 call to your immune system, your cells start listening to and responding to other voices and directions. It's like an old-fashioned party line—too many callers talking so the wrong messages—or no messages at all—get through.[16]

Your cells become unglued.

Provocative new research has uncovered one way cellular transmission is interrupted. Studies have found that even low-level EMFs may rupture delicate cell membranes, releasing calcium from cells as well as changing the way calcium ions—electrically charged calcium atoms—bind to the surface of the membrane. For example, Adey and his colleagues found that exposing newly hatched chicks to a 16 Hz frequency caused their brain cells to leak calcium ions.[17] Since calcium ions are the glue that holds together cell membranes, which are only two molecules thick, the membranes are likely to weaken and tear, allowing toxins to enter and contents to spill out. They literally become unglued.[18]

Obviously your cells need some calcium. There's even a natural system in place to make sure they get the right dose. But what happens when there's a flood of calcium ions from a torn membrane in the main part of the cell? It depends on what your cells are doing at the time. If you're sick or injured, your cells are in the process of healing you, so these extra calcium ions help speed the process.[19]

When calcium ions pour into one or more of your one hundred billion brain cells, which use calcium in small doses to make neurotransmitters, they may release those chemical messengers too soon, too often, or at the wrong time, creating false messages that tell you that you're in pain or bring on neurological symptoms, such as headaches, an altered sense of taste or smell, tingling, or numbness. Those are just a few of the symptoms experienced by the people we just met and part of a collection of problems experienced by people who are hypersensitive to EMFs.[20]

Too many calcium ions in your brain cells may also impair your life-saving ability to assess a situation correctly—like when you're at the wheel of a car. Noted British scientist Andrew Goldsworthy, Ph.D., honorary lecturer at Imperial College of London, suspects that the increase in accidents among cell phone users (in one in four crashes, a driver is on a call) has less to do with distraction than with delayed response caused by the flood of calcium ions into brain cells. This flood creates what he calls "a mental fog" of false information, obscuring the ability to react to, say, a child on a bike pulling out between two cars or a deer bounding from the woods at twilight. After all, we're often distracted at the wheel when we're talking with a passenger, listening to a radio talk show, or engrossed in an audio book, none of which have been linked to increased accident risk. There's obviously something more, something physical related to phone use.[21]

And in fact, a study of young adults aged twelve to fourteen in Australia found that those who used their cell phones the most suffered from poor memory and delayed reaction time—yes, even when they weren't on the phone.[22]

Chemicals pouring from your ruptured cells damage your cellular DNA.

Our bodies have an amazing defense system. Just as cell membranes offer some protection from EMFs (though not enough), a healthy cell membrane will also self-heal. But, before it repairs the tear, it may release a digestive enzyme called DNAase, which can destroy or damage DNA, potentially turning your genetic material into a precursor to disease by altering its important directions on how and when to grow, divide, and die. Studies using cell phone signals have found evidence of just that effect. For instance, in one Greek study of fruit flies, whose short life span makes them the perfect subject for basic genetic research, researchers found that exposure to mobile phone signals for only six minutes a day for six days actually fragmented the genetic material in the cells that produced the flies' eggs, and half of the eggs died.[23]

EMFs may disrupt normal cell division.

Electromagnetic fields may strike at cellular DNA in other ways too. Scientific research has found that exposure to ELFs, for example, speeds up cell division and reproduction. During the cell division process, known as mitosis, DNA is reproduced, chromosomes line up in pairs, and then pull apart to create a daughter cell that should be the spitting image of its mother. Exposing cells to ELF disrupts that orderly process of chromosome matching and detaching, so that the two new cells don't get equal amounts of the genetic information. This can result in scrambled messages. The consequence? Damage to fertility or a developing fetus.[24]

EMFs create oxidative stress that further damages DNA and other physical processes.

Evidence from animal studies suggests that exposure to the level of electromagnetic force that's produced by something as mundane as your refrigerator may create free radicals, unpredictable molecules whose unpaired electrons seek to attach themselves to electrons in other molecules. Those most fundamental things we do—breathing and eating—cause our body to react with oxygen. It's perfectly normal. But it's a process that can go awry. When metal becomes oxidized, for example, it develops rust. When you slice open an apple and leave it in the air, it turns brown. Oxidation can turn fats rancid, which is why you're advised to keep oils tightly sealed and in a cool, dark place. The same thing happens in your body. Fat becomes rancid—scientists call it lipid peroxidation, and it sets your cardiovascular system up for the buildup of hardened lumps of fat and other debris on your arteries. Those clogs can cause heart attacks and strokes. Free radicals contribute to arthritis by oxidizing joint fluid, making it less lubricating. They can cause DNA damage to your cells, making cell membranes so rigid that nutrients can't get in and ultimately make the cell so fragile it breaks, allowing toxins to come in and fluid to drain out before it finally collapses. This process is considered the root cause of aging and disease, from cancer to Alzheimer's. At the University of Washington, scientists Henry Lai and Narendra P. Singh found that free radical creation at a 60 Hz alternating current—typically found in homes that have no wiring problems and are not located near power lines—caused breaks in the DNA of brain cells of rats exposed for *only twenty-four to forty-eight hours*.[25] The rest of us are exposed 24/7.

SCIENCE TAKES NOTICE

The evidence is accumulating. One of the first studies linking magnetic fields from power lines to adverse effects on human health was published in 1979 by two Denver researchers, the late Nancy Wertheimer, Ph.D., and physicist Ed Leeper. Based on Wertheimer's field studies of childhood cancers in the Denver-Boulder area, the two reported that children who lived one or two houses from what are called step-down transformers (the barrel-shaped devices mounted on the power poles in your neighborhood) had a two- to three-fold increase in childhood cancers, specifically leukemia and brain tumors.[26] In 1986, a similar study conducted at the University of North Carolina, Chapel Hill, confirmed their findings.[27]

Although a number of studies done since have disputed the leukemia-EMF link, there have been at least thirty not only confirming the original 1979 work, but expanding on it to associate transmission power lines, hair dryers, common household appliances, video games, and microwave ovens to children's cancers.

In fact, David Carpenter, M.D., dean of the School of Public Health at the State University of New York, has been quoted as saying that he believes up to 30 percent of childhood cancers stem from EMF exposure.[28] And it doesn't take much. In several of these studies, the risk was elevated when children lived near magnetic fields that were one thousand times *lower* than the existing safe exposure limit established by the International Commission on Non-Ionizing Radiation Protection.

Since the Leeper-Wertheimer study, hundreds of studies have found that exposure to magnetic fields (EMFs) may be associated with a variety of conditions, including Alzheimer's disease, heart disease, amyotrophic lateral sclerosis (ALS, or Lou Gehrig's disease), heart disease, miscarriage, birth defects, infertility, and mood disturbances such as depression.

To add to this, some experts are now saying that proliferating technology is overburdening our aging electrical infrastructure, exposing us to high frequency EMFs in our homes, offices, and schools. Surges of high frequency voltage or electromagnetic radiation from radio waves, currents that run along grounded lines or water pipes, or high frequency spikes and harmonics (distortions in the current or wave) from our appliances and other electronic sources are contaminating the low frequency lines, creating a hybrid now being called dirty electricity.

Studies suggest these "freaky" frequencies may be the cause of sick building syndrome—a constellation of symptoms including headaches, allergies, fatigue, skin irritation, depressed mood, and disruptive behavior in children—and some cases of attention deficit disorder (ADD). There's also evidence that it may raise blood sugar in diabetics and increase symptoms in those with multiple sclerosis.

THE LIST GOES ON

In just the past five years, new research has painted a more detailed picture of the health and environmental effects of electromagnetic pollution. Here are just a few highlights from the hundreds of studies I've reviewed:

- In 2006, a study of the cell phone habits of nine hundred people with brain tumors, conducted by the Swedish National Institute for Working Life, found that those who used cell phones for two thousand cumulative hours had a 240 percent increased risk for a malignant tumor on the side of the head where they usually held the phone.[29] Two years later, Israeli researchers found that those people who kept their cell phone against one side of their head for several hours a day were 50 percent more likely to develop a rare salivary gland tumor on that side (just like mine).[30]

- A study published in the journal *Epidemiology* in July 2008 reported that children born to mothers who used cell phones while pregnant and whose children used cell phones by age seven were 80 percent more likely to be hyperactive and to have emotional and behavioral problems.[31]

- Many studies have found that EMFs can interfere in the body's nighttime production of the hormone melatonin, which is vitally important for sleep and the lack of which can impair immunity.[32]

- Other studies and personal reports link even minimal EMF exposure to sleep disturbances, immune-system suppression, brain wave changes, headaches, light sensitivity, heart arrhythmias, chronic fatigue, memory problems, ringing in the ears, depression—associated with a new problem, called electrosensitivity.

The Price of Ignoring Early Warnings

If you grew up in the 1940s and '50s, a trip to the shoe store was a treat because you got to see inside your feet, thanks to that great little toy, the pedascope. Shoe stores had them to keep kids occupied while their parents shopped, and because it was a fluoroscope, kids were exposed to a hefty dose of radiation while they stayed out of mom's hair.

This is just one of the dangers we've experienced—and continue to face—when the real impact of technology is either unknown or ignored. In the 1930s and '40s, women who wanted to have undesirable hair removed were exposed to X-rays, as were children with ringworm. In the '30s, mental patients were treated with radium, despite the fact that exposure to the element eventually killed Marie Curie, who, with husband Pierre, was awarded the 1903 Nobel Prize in Physics for discovering it. Likewise, Thomas Edison's assistant Clarence Dally's arm was amputated and he died as the result of exposure to X-rays in 1904. In fact, it was the early, damaging effects of radiation that first triggered scientists to suspect it might also be used to kill cancer cells.[33]

The rise of regulating agencies during the post–World War II period was the first step in developing safety standards (which, of course, are only necessary once you finally acknowledge that something is dangerous). A study by an epidemiologist in the late 1950s that found that exposing fetuses to X-rays was linked to leukemia in children was at first disbelieved, but once the study was replicated time and again, it provided evidence for the current no–X-ray policy by obstetricians. However, the early study wasn't acted on until much later, resulting in one estimate that 5 percent of childhood cancers were caused by prenatal X-ray exposure.[34]

Radiation expert Barrie Lambert, Ph.D., who details this history in a paper in the journal of the European Environment Agency, called, aptly, "Radiation: Early Warnings, Late Effects," writes: "It could be claimed that these numbers of leukemias would have been saved had the work [of the epidemiologist] been acted on earlier. A similar and contemporary story may be unfolding in relation to the childhood leukemia risk ... in the United States" precipitated by exposure to EMFs.[35]

- There has also been some evidence that EMFs may contribute to some of the leading environmental issues of our time. German studies suggest that destruction of forests (in Germany and in the western United States) once blamed on acid rain may actually be the result of constant bombardment from 60 Hz power lines and the RF waves from communications equipment. And some researchers suspect that electropollution may be in part responsible for the changing weather patterns now blamed on global warming.[36]

THE EXCEPTION THAT PROVES THE RULE

To be perfectly honest, many scientists still regard the low-level fields and the high-level radio waves to which we're exposed to be entirely benign. They argue that these are fairly weak fields that diminish rapidly the farther away you get from them. And many studies have looked at the same data and have not shown the same effects.

A group of top international scientists, writing in the 2007 *BioInitiative Report,* which called for further examination of EMFs and their effect on public health, were clear: conflicting studies should not be taken as an "all clear." In fact, they note, "there should be *no effect at all* if it were true that EMF is too weak to cause damage." And if EMFs had no substantial effect on the human body, the report says, then their use as *therapeutic tools* would be little more than quackery.[37]

It's true. Mainstream medicine harnesses the power of EMFs to heal. It seems like the ultimate paradox. Broken bones are mended and wounds healed by pulsed EMF stimulation; pain is eased by transcutaneous electrical nerve stimulation (TENS), the application of electricity that appears to activate the body's own pain-relieving (opioid) system; unsightly port wine stains are removed by laser, a light source of EMF; and depression is lifted by transcranial magnetic stimulation, which uses weak electrical currents and rapidly varying magnetic fields to excite brain nerve cells.

Currently being tested on humans is the use of low-intensity EMFs to literally jiggle the electrically charged particles in cells hundreds of thousands of times per second, which can disrupt the division of cancer cells, preventing them from spreading. Israeli researchers studied ten people with glioblastoma multiforme—the same deadly form of brain cancer that killed Senator Edward Kennedy. Those who received the low-intensity EMF therapy lived longer—median survival rate: sixty-two weeks—than most people with the disease, which usually results in death within twelve months of diagnosis.[38]

And yet, in the midst of all this gloom and doom, there are real glimmers of hope and true healing, as you will find out shortly. There are positive ways to reduce the effects and negative influences of EMFs without giving up the comfort and conveniences of modern-day living.

LIVING WITH TECHNOLOGY

At this point, you may be doing what I did while researching this book—making a mental list of every electronic device you own and assessing which ones you could live without, or maybe you're planning your move to a relatively remote part of the globe. The truth is that some technologies do present a clear and present danger. There are hundreds of reliable independent studies that say so. But there are also ways of mitigating their effects safely and practically—which is what this book is all about. We may not know the exact extent to which EMFs pose a threat, but we do know the ways in which the technologies that generate them benefit us. So living safely with technology is definitely a balancing act. You will have to intervene in your own living situation. You don't have to do without electric lights, satellite TV, your microwave, cell phone, or BlackBerry as long as you are reducing exposure, prudently avoiding overuse, and implementing some of the cutting-edge and grounding lifestyle therapies you will be introduced to later in the book. After all, even if you get rid of every single electric gizmo and appliance in your house, you will still be surrounded by them for hours every day as you spend time at work, in your car, and in public places like restaurants, theaters, and malls.

The purpose of this book is to show you the best way to live with our technology so you can prevent and even reverse its negative side effects. I will empower you with real-life information so you can start making safer lifestyle choices. It's not a problem that is easily or quickly fixed but it is a problem that each of us can begin addressing in some way in our own lives and almost immediately to protect the well-being of our precious bodies and that of our loved ones. Here's what you'll learn in each of the chapters that follow:

In the next chapter, "Lifting the Veil," you'll pick up the first clues to the mystery of the epidemic of chronic stress, sleep deprivation, and cancer in this country. You'll be surprised to learn what all those modern-day ills have in common and why EMFs are a likely coconspirator.

Then, you will take a quiz in chapter 4, "How Zapped Are You?" This simple quiz will help you to determine where, when, and how often you are exposed to electropollution in key aspects of your life. Chapter 5 zeroes in on your home environment and will enable you to implement an action plan as you "Zap-Proof Your Home." In chapter 6, you will learn about several amazingly inexpensive devices—one of which you may already have on hand—which you can use immediately to start your own investigation as you transition into "Advanced Zap-Proofing." I am going to take you through all the potential "hot spots" inside and outside of your home where you may want to use even more sophisticated meters to measure the actual strength and location of harmful EMFs. Hint: Don't forget to measure the field surrounding the water pipes; it may be the highest source of EMFs in your entire environment.

How harmful is dirty electricity? I'll tell you all about it in chapter 7, "The Newest Zapper: Dirty Electricity" as well as introduce you to Catherine Kleiber, now a crusading educator about the perils of electropollution, who fought for years to discover the cause of the ailments that befell her when she moved to a Wisconsin farm with her new husband. Most importantly, you'll gain some valuable insights about a fast, simple, science-tested solution that can remove much of the dirty electricity in your home.

In chapter 8, "Zap-Proof Your Phone," you'll find dozens of well-researched tips on using your cell phone in the safest possible way as well as vital information on choosing a cell phone. Chapter 9, "Zap-Proof Your Kids," is a must-read. Children, by virtue of their size and their maturing brains and bodies, are most vulnerable to the effects of EMF exposure. Studies have also linked EMFs to fertility problems. Here, you'll find instructions on how to protect fertility and your children—even some advice on how to approach the topic of limiting cell phone use by teens (which I'm sure you don't think is possible!).

In chapter 10, "Zap-Proof Your Work Environment," you'll meet the electrical engineer who worked at a naval facility and couldn't get through the day without napping—until his employer did something far beyond the call of duty for him.

In chapter 11, "Other Zappers in Your Life," you will find out which medical tests you should be wary of and why. You will learn what parts of your body are most vulnerable to electromagnetic radiation. I'll also reveal the forgotten dangers from nuclear power plants and dump sites near your neighborhood as well as radon in your home.

Being electrosensitive myself, I'm always on the lookout for ways to keep healthy in light of the most current environmental assaults. As a practicing nutritionist for nearly three decades, I know the value of eating to protect yourself from the slings and arrows of modern life, including stress, infection, disease, and pollution—even electropollution. In chapter 12, you'll find my "Zap-Proof Superfoods and Seasonings," plus some out-of-this-world recipes to help you get the most EMF protective and restorative foods deliciously into your diet. Chapter 13 provides an extra dose of healthy prevention and dietary insurance in my "Zap-Proof Minerals and Supplements."

In the resources section, I'm going to share with you the go-to websites for the latest EMF info, updates, alerts, and protective products of all kinds. You will also find a book list and energy medicine healing devices and formulas to combat and perhaps even reverse the detrimental side effects of electropollution. You'll learn how to get zapped in a therapeutic way to counteract the EMFs that are making so many of us sick.

In the meantime, let's shift our focus and talk about the roles that certain hormones and stealth stress play in our overall health.

LIFTING
THE VEIL

You may not be able to see electropollution, but your body responds to it as though it were a cloud of toxic chemicals or a glass of water contaminated with heavy metal. Invisible or not, poison is poison.

We've already seen the various ways it can damage your cell membranes and create reactive molecules called free radicals that cause so much damage to our bodies that most scientists now think they're the cause of everything from Alzheimer's to aging and premature death. But new research suggests it's more insidious than we could have imagined: this invisible pollution not only creates free radical damage, it reduces the body's ability to heal itself from that damage.

It does it in part by suppressing the body's production of important antioxidants. One of those is melatonin, which is also a hormone that regulates the sleep-wake cycle.

Fortunately, as you'll learn in later chapters, you're not defenseless. Lifestyle changes—including a special immune-boosting diet I've developed—will help you protect yourself from electropollution.

THE MELATONIN CONNECTION

Dozens of studies have found that even low levels of EMFs can depress the body's production of melatonin. Created by the pineal gland, a small pinecone-shaped gland the size of a pea buried deep inside the brain, melatonin is more than just your body's own alarm clock, linked to insomnia and jet lag. Called the Dracula hormone because its production peaks at night, its role in protecting us from disease was first noted when studies began finding that night workers experienced higher rates of breast cancer than those who held down nine-to-five jobs. Now we know that the hormone can increase the effectiveness of the body's own killer cells called lymphocytes, which can fight off the foreign invaders, like mutated cells, it encounters. It doesn't just help you sleep—it potentially saves your life. This is one remarkable chemical.

THE IMMUNE SYSTEM WARRIOR

Melatonin bolsters your immune system in another critical way. It increases the antioxidant activity of two powerful chemicals, superoxide dismutase (SOD) and glutathione peroxidase, that occur naturally. SOD, which is also an anti-inflammatory, helps repair cells, specifically the damage they incur from the most common free radical in the body—superoxide. Recent studies suggest a link between low levels of SOD and amyotrophic lateral sclerosis (ALS), also known as Lou Gehrig's disease, a fatal condition of the nerve cells of the brain and spinal cord, leading to total paralysis and eventually death.[1] Significantly, research has traced a connection between ALS and occupational exposure to EMFs.[2] One study, done by the Danish Cancer Society, found a clear association to this disorder. The missing link might be melatonin.[3]

Much like SOD, glutathione is a powerful antioxidant and detoxifier that occurs in every cell in your body. Like a resident handyman, it can repair any free radical damage on the spot as well as clean up any toxins and the injury they cause. It's interesting to note that in people with cancer, AIDS, and other serious diseases, glutathione levels are severely diminished, suggesting the enzyme plays a major role in immune system defense.[4] In fact, researchers have been testing the use of glutathione in treating certain cancers. Like most antioxidants, glutathione works by supplying an extra electron to unpaired free radical molecules, returning them to a benign state. But in giving up an electron, the glutathione itself becomes a free radical. What melatonin does is help glutathione regain its antioxidant status—effectively helping it live to fight another day.

MELATONIN—METABOLIC ENHANCER

The pineal gland is also a supplemental source of thyrotropin-releasing hormone (TRH), which the body uses to create thyroid-stimulating hormone (TSH). The antioxidant hormone melatonin also stimulates TSH, creating thyroid hormones that control your metabolism.

It's long been known that radiation zeroes in on the thyroid gland as though it were painted with a target. The reason: the thyroid needs iodine from the bloodstream to produce the hormones that regulate your energy and metabolism. But it can't distinguish between normal iodine and radioactive iodine. It will absorb whatever is there, which is why every good radiation protection kit contains iodine tablets. After the 1986 nuclear accident at Chernobyl, exposed children started developing thyroid cancer sooner and in larger numbers than scientists expected. The rate of thyroid cancer rose an astonishing 2400 percent! But in nearby Poland, where government medical officials distributed potassium iodide pills to protect residents from the airborne radiation from the meltdown, there was no corresponding rise in thyroid cancers.[5]

There's evidence that even low-level radiation may have an effect on the thyroid. Studies in rats have found that exposure to both 50 Hz (European electrical system) and 900 MHz (cell phones) EMFs decrease the production of the thyroid hormone, a marker of hypothyroidism, which can cause fatigue, cold intolerance, depression, muscle cramps, joint pain, weight gain, low heart rate, and constipation.[6] It can also adversely affect your cholesterol, elevating levels of bad cholesterol—LDL—that can raise your risk of heart disease.

Canadian researchers found that people who had been exposed to radiation through the environment, the workplace, or treatments for acne or other conditions at least three years before they had surgery for thyroid cancer were more likely to have a more aggressive form of the disease that recurred or metastasized. The average age at first exposure to radiation was 19.4 years, and their cancers were diagnosed an average of 28.7 years later. In the past twenty years, the incidence of thyroid cancer has jumped dramatically, as has the incidence of cancers that have spread.[7]

MELATONIN—POTENT ANTIOXIDANT

Melatonin is also one of the most highly potent antioxidants itself—five times more powerful than vitamin C—which protect cells from DNA damage caused by free radicals. There is evidence that melatonin reduces the damage to blood vessels that leads to atherosclerosis (hardening of the arteries), lowers cholesterol, and may help reduce risks of neurodegenerative diseases like Alzheimer's by preventing the death of neurons (nerve cells that help carry messages throughout the body's electrochemical communication system).[8]

LOW MELATONIN LINKS TO
FREE RADICAL DAMAGE

You can see the important role melatonin plays in the body, particularly in the immune system. But what happens when melatonin production slows, as it does when we age and when we're exposed to EMFs? Many scientists now suspect that the low melatonin levels we experience when we're over sixty may explain, at least in part, age-related insomnia and the increase in our vulnerability to disease. The incidence of many degenerative and killer diseases rises dramatically as we head into our senior years. With less circulating melatonin, the body is more vulnerable to free radical damage which, of course, is linked to those very diseases.

Since the hormone also beefs up the power of our immune system's lymphocytes to eliminate any cancer cells that slip by our other defenses, having less of it in circulation takes out another battalion of our disease-battling army. Not surprising then, low melatonin levels have been implicated not only in breast cancer, but in melanoma—the deadliest form of skin cancer—and malignancies of the ovary and prostate. Low levels are also a factor in Alzheimer's and Parkinson's diseases, both of which are more likely to occur in the elderly.[9]

There's persuasive evidence that, in both animals and humans, EMFs not far above the frequency of earth's ambient EMF alter melatonin production.[10] A German study focusing on residents who lived near a recently installed mobile phone tower detected lower levels of both melatonin and serotonin in blood samples taken from the participants when compared to samples that had been taken before the tower was operational. In fact, 84 percent of the study group had "massive" decreases in serotonin and, not surprising, nearly all reported symptoms including depression, lethargy, listlessness, lack of appetite, and agitation. About 54 percent had lower levels of nighttime melatonin and suffered from sleep disturbances—most woke up between two and four A.M. and had trouble getting back to sleep, problems similar to those reported by older people with naturally declining melatonin. They also reported feeling tired and unable to concentrate the next day.[11]

All those symptoms were described by Soviet scientists in the 1950s as signs of "radio wave sickness," a condition found mainly in radio and radar operators exposed to RF, or microwave, fields produced by their equipment.[12]

Not only that, but other studies have found that both animal and human blood cells exposed to both RFs and ELFs showed a decrease in the activity of SOD. In the case of the human blood cell research, the hampered activity was seen after *only one minute of exposure!*[13]

EMFS—THE TRIPLE THREAT

What this means, as I've pointed out, is that EMFs are not only creating DNA-damaging free radicals but at the same time are also suppressing the production of the body's defense systems. First, they slash back the über-antioxidant hormone melatonin. That in turn reduces the levels and free radical scavenging power of the antioxidant enzymes SOD and glutathione. In other words, exposure to EMFs effectively slays your immune system's border guards while at the same time sending in fresh enemy reinforcements.

We know from studies done at the Cancer Therapy & Research Center in San Antonio, Texas, that our body's defenses *are* weaker after exposure to EMFs. Human cancer cells grow twenty-four times faster than cells that haven't been exposed to EMFs and show "greatly increased resistance to destruction by the cells of the body's defense system."[14]

EMFS CAUSE SUBLIMINAL, OR STEALTH, STRESS

There is far from consensus on how EMFs affect the human body. Studies have been inconclusive; scientists with equally sterling degrees disagree, sometimes angrily. But if we pay attention to the wisdom of the body, a clearer picture emerges. While the researchers squabble, your body is reacting to EMFs as though they were public enemy number one. Every time you're chatting on your cell phone, your body is producing heat stress proteins. This response is usually a signal that the body is in heat shock or exposed to toxic chemicals, heavy metals, or other environmental hazards. *It means your cells are in distress,* even if you feel perfectly fine.

Test animals exposed to electric and magnetic fields at levels to which we're exposed show a wide range of stress reactions. They have higher levels of adrenaline, which is part of the body's fight-or-flight chemical response to danger; they produce more stress hormones such as cortisol; their immune systems are suppressed; and their heart rate and blood pressures rise. Yet the animals don't appear stressed.[15]

Of course, stress doesn't just involve that emotional response we all know and recognize. When you're stressed, your entire body becomes involved in a massive chemical mobilization to help protect you from whatever is threatening you. It happens in virtually an instant, but it is as complicated a battle strategy as D-Day. Here's what happens:

- The flow of oxygen and blood is shut down to all but the major organs like the heart and brain.

- Any systems—including immunity and digestion—that aren't necessary to help you fight your enemy or run away from danger are depressed.

- To pump you up for the coming battle or flight, a chemical flood increases your heart and breathing rate, ups your blood pressure, infuses your bloodstream with energy in the form of glucose, also known as sugar, and increases blood flow in your arms and legs by dilating your blood vessels.

EMFs seem to trigger what preeminent researcher Robert O. Becker, M.D., called "subliminal stress"—stress that's under your intellectual radar but picked up with acuity by your body's internal antenna.[16]

Your body knows instinctively that EMFs pose a threat, and it responds accordingly. Naval studies that looked at this phenomenon in animals found that exposure led to an increase in a neurotransmitter called acetylcholine in the brain stem. Acetylcholine is the primary neurotransmitter in the central nervous system, and it's a little like the person in the lifeboat who stands up and starts screaming hysterically (before someone wisely slaps him). It causes excitability and arousal, so theoretically that chemical flood sounds an alarm in the animal's body even though the animal never detected anything harmful or felt stressed.[17]

But what if the danger never goes away? As we've seen, the world we live in is bathed in EMFs of every frequency, and most of them don't turn off. Science is clear on what happens when we're exposed to chronic stress, when that chemical warning system never shuts down. It works exactly like free radical damage—attacking every organ system.

Your Heart under Stress

Constant increases in heart rate, elevated levels of stress hormones like adrenaline and cortisol, and rising blood pressure set the stage for hypertension, which can lead to heart attack or stroke. Stress can also raise cholesterol levels and boost inflammation, both of which are linked to cardiovascular disease. Many of these symptoms have also been linked to EMF exposure, particularly in people like those we met in chapter 2.

Your Endocrine System under Stress

Under stress, your hypothalamus (brain region regulating body temperature, hunger, thirst, fatigue, and circadian cycles) signals your autonomic nervous system to produce the stress hormones adrenaline and cortisol to give you the power to fight or flee—and that triggers a release of sugar (glucose) into your bloodstream to give you an even greater energy boost. Chronic stress can keep that system working, leaving you vulnerable to diabetes, which is characterized by elevated blood sugar. Studies and anecdotal reports confirm a rise in blood sugar when you're exposed to EMFs, particularly the hybrid low and medium frequency variety known as dirty electricity and magnetic fields.

Your Digestive System under Stress

When you're under chronic stress, your body produces a number of other chemicals that may literally hit you in the gut: cytokines, components of your immune system that cause inflammation; serotonin, a neurotransmitter that affects smooth muscle contraction; and the enzyme protease that regulates protein digestion. In some people, these chemicals may slow the digestive process, so they get bloating, pain, and constipation, while others may experience diarrhea, often soon after eating. Stress can also increase chronic heartburn and acid reflux, which are risk factors for esophageal cancer. Digestive disorders are one of the many symptoms of a condition called electrosensitivity syndrome, which affects people with a variety of often debilitating symptoms when they're exposed to even low-level radiation.

Your Reproductive System under Stress

The production of stress chemicals can affect a man's testosterone and sperm production and maturation, and may cause erectile dysfunction or impotence. In women, it may reduce blood flow to the reproductive organs and interfere with proteins in the uterine lining that are involved with implantation of a fertilized egg. Some fertility specialists believe that chronic stress is responsible for as much as 30 percent of all fertility problems. EMF exposure has been linked to poor sperm quality, fetal malformation, birth defects, and miscarriage.[18]

Your body goes on red alert when it is exposed to EMFs, so you need to become very vigilant and mount a powerful defense. That's why you really need to get a true assessment of how exposed you really are. That's where the quiz in the next chapter comes into play.

HOW ZAPPED
ARE YOU?

You may be surprised—even shocked—to discover that on a daily basis you are exposed to some form of electromagnetic radiation that may be compromising your health. Up until now, you probably have not been aware of this hidden form of pollution and how zapped you are in your home, at work, and even where you play. And, if you work in certain occupations, live or play within a certain proximity of broadcast towers, cell towers, or electrical power lines or transformers, or have specific lifestyle habits, your zapped load may be seriously higher than the norm, which might just explain a whole variety of seemingly unrelated or unresolved symptoms you haven't been able to figure out.

My purpose is to help you take a realistic inventory of your daily exposure to electropollution so you can start to identify and focus on the changes and choices you need to make to enhance your well-being. And while this quiz is by no means a scientific questionnaire, it will provide you with a level of insight into the potential health challenges associated with your individual exposure, which in turn will help you decide what positive lifestyle changes you want to make—and to what degree—for you, your family, your associates, and the planet in general.

This can be such an eye-opener that I encourage you to take the quiz with your partner and kids. You will note that each of the quiz items is followed by a number that indicates its potential degree of importance as a risk factor if the answer is yes. While everyone is exposed, in this day and age, it may give you pause to reflect to add up the factors that specifically apply to you and your family. In general, the higher your number the greater the risk of electropollution and its consequences. Total each section to see where you have the highest risk of being zapped. That being said, if you are already experiencing a number of puzzling symptoms, much like the people in chapter 2, additional exposure, even if it is slight, can create toxic overload and make you electrosensitive.

You'll notice I also include a "don't know" checkbox. I hope you will take some time to follow up on the questions "you don't know." Your current and future health just might depend upon the answers you find, because what you don't know can still hurt you.

Quiz

At Home

The Location of Your House

Are there above-ground local power distribution lines within one hundred feet of your home? (2)

☐ Yes   ☐ No   ☐ Don't Know

Are there transformers within fifteen feet of your home? (2)

☐ Yes   ☐ No   ☐ Don't Know

Are there major transmission lines within a half mile of your home? (2)

☐ Yes ☐ No ☐ Don't Know

What the dangers are: Living near power lines that produce EMFs has been linked to serious diseases such as leukemia in children.

Potential remedies: See chapter 6.

Equipment Throughout the House

When you used a Gauss meter to measure the magnetic fields in areas of your house where people spend many hours per day, did the readings exceed 1 mG even three feet away from your electronic gear, appliances, and wiring? (3)

☐ Yes ☐ No ☐ Don't Know

What the dangers are: Above 1 mG, EMFs are considered biologically active, meaning they can affect the physical processes in our bodies, including interrupting cell communication, damaging cellular DNA, and potentially causing us to become ultra-sensitive to even low level nonionizing radiation.

Potential remedies: In chapters 5 through 7, you'll learn how to measure your home EMFs and all the steps to reduce the low-level radiation in your home to 1 mG or below.

Do you use cordless phones? (1)

☐ Yes ☐ No ☐ Don't Know

What the dangers are: The base station of the cordless phone is constantly transmitting RF waves—even when it is not in use.

Potential remedies: See chapter 5.

Do you use cell phones or PDAs? (2)

☐ Yes ☐ No ☐ Don't Know

What the dangers are: Studies indicate that cell phone use, particularly if you've been using one for a decade or more, is linked to brain tumors. Wireless personal digital assistants produce additional EMFs from the batteries, which can be significant.

Potential remedies: See chapter 8.

Do you have dimmer switches anywhere in the house? (2)

☐ Yes   ☐ No   ☐ Don't Know

What the dangers are: Dimmer switches are a surprisingly major source of dirty electricity—and create very large EMFs.

Potential remedies: See chapter 6.

Is your home computer system set up for a wireless Internet connection? (3)

☐ Yes   ☐ No   ☐ Don't Know

What the dangers are: A wireless system is on 24/7, emitting low-level radiation and radio waves throughout your house. The closer you are to the source, the more radiation you will be exposed to. You should also know that you can be exposed from your neighbors' wireless connections next door or even across the street. It's important to know this so you can measure and shield accordingly.

Potential remedies: See chapter 5.

If you are a male, do you carry your cell phone turned on and in your pocket? (1)

☐ Yes   ☐ No   ☐ Don't Know

What the dangers are: Studies have found lower sperm quality in men who carry their switched on cell phones in their pockets.

Potential remedies: See chapter 8.

Is there a refrigerator, TV, stereo, microwave oven, cell phone charger, black box power supply for electronics, or other appliance on the other side of the wall from the sofa where you watch TV? (1)

☐ Yes   ☐ No   ☐ Don't Know

What the dangers are: Your refrigerator creates one of the largest magnetic fields in your house because of its motor and fan. You may not spend a lot of time in the kitchen, but if your living room or family room or a bedroom is on the other side of the wall from your refrigerator, you and your family may be lounging in that field for hours. The other sources of magnetic fields mentioned above produce smaller fields, but they are still significant when you are within three feet of the device.

Potential remedies: See chapter 7.

Do you use electrical tools in your home workshop or in making home repairs throughout the house? (1)

☐ Yes   ☐ No   ☐ Don't Know

What the dangers are: Even the smallest electrical tools produce a hefty EMF. A power saw can produce a field of 1,000 mG at the distance at which you hold it.

Potential remedies: See chapter 5.

The Bedroom

Do you use an electric blanket? (2)

☐ Yes   ☐ No   ☐ Don't Know

What the dangers are: These blankets can produce very large AC electric and magnetic fields and have been linked to infertility.

Potential remedies: See chapter 5.

Do you sleep in a waterbed with a heater? (2)

☐ Yes   ☐ No   ☐ Don't Know

What the dangers are: A large AC magnetic field emanates from a water bed heater when it's plugged in.

Potential remedies: See chapter 5.

Do you have a TV or sound machine in your bedroom that is always plugged in and is closer than three feet to your bed? (1)

☐ Yes   ☐ No   ☐ Don't Know

What the dangers are: AC magnetic fields that emanate from a television or sound machine can interfere with the production of the hormone melatonin—which governs your circadian rhythms including your ability to fall asleep and also boosts your immune system and is a powerful antioxidant. Your body manufactures most of the melatonin you need at night, so you're particularly vulnerable to the effect of AC magnetic fields in the room where you sleep.

Potential remedies: See chapter 5.

Do you have computer equipment in your bedroom? (3)

☐ Yes   ☐ No   ☐ Don't Know

What the dangers are: Modern computers emit a wide range of EMFs at frequencies from 60 Hz to millions of Hertz. Your bedroom should be as technology-free as possible, not only to help you sleep but to protect you from the conditions that are affected by low levels of melatonin.

Potential remedies: See chapter 5.

Do you sleep with your head near a wall that contains wiring or a power outlet? (1)

☐ Yes   ☐ No   ☐ Don't Know

What the dangers are: The AC electric fields from your wiring produce an unnatural AC voltage on your body that has been observed to cause numerous physiological problems.

Potential remedies: See chapter 5.

Is there a TV, computer, or large appliance (such as a refrigerator) on the other side of the wall from your bed? (1)

☐ Yes   ☐ No   ☐ Don't Know

What the dangers are: Walls are no barrier to AC magnetic fields, and often you're exposed to fields because the risk actually comes from the appliances emitting AC magnetic and electric fields from the other side of the wall.

Potential remedies: See chapter 5.

Is there a fluorescent or halogen light or a fan in the ceiling of the floor below your bedroom? (1)

☐ Yes   ☐ No   ☐ Don't Know

What the dangers are: The AC magnetic fields produced by lights and ceiling fans are actually greater in the rooms above them because you are closer to the source.

Potential remedies: See chapter 5.

The Kitchen

Do you have a microwave oven? (1)

☐ Yes   ☐ No   ☐ Don't Know

What the dangers are: Microwave ovens leak RF even when they're new. RF goes through the screen that you look through to see your food and right into your eyes. This is allowed by outdated regulations. Opening and closing your microwave over time can erode the tight seal on the door and can allow more dangerous radiation to leak out.

Potential remedies: See chapter 5.

Do you eat in your kitchen? (1)

☐ Yes   ☐ No   ☐ Don't Know

What the dangers are: If you are within three feet of your appliances, you may be exposed to electromagnetic fields—potentially large ones produced by your refrigerator.

Potential remedies: See chapter 5.

The Bathroom

Do you use an electric toothbrush? (1)

☐ Yes   ☐ No   ☐ Don't Know

What the dangers are: It's great for your teeth, but the charging base of an electric toothbrush creates magnetic fields. The toothbrush itself produces a 260 Hz field, and you may be holding it in your mouth for two minutes, if you're brushing correctly.

Potential remedies: See chapter 5.

Do you use an electric razor? (1)

☐ Yes   ☐ No   ☐ Don't Know

What the dangers are: The Electric Power Research Institute (EPRI) measured the magnetic fields around common household electrical devices; one of the eye-opening findings was the electric razor. It produced a 600 mG field at six inches away.

Potential remedies: See chapter 5.

Do you use a hair dryer? (2)

☐ Yes   ☐ No   ☐ Don't Know

What the dangers are: The average handheld hair dryer produces a 25 mG field around the motor. At your hand and around your head, that would be as much as a 50 mG field, which is probably more than you're exposed to by your neighborhood power lines. But, because you don't use hair dryers for extensive periods of time—even at close proximity—this exposure is not as potentially serious as others. That said, the frequent use of hair dryers on children's heads is more dangerous because their smaller heads and thinner skulls allow greater magnetic field penetration than occurs in adults.

Potential remedies: See chapter 5.

Total: _____

At Work

Do you spend more than half of your day within approximately three feet of large electric motors, generators, or the power supply in a building? (3)

☐ Yes ☐ No ☐ Don't Know

What the dangers are: Motors, generators, and electrical supply equipment can produce extremely large AC magnetic fields and have been linked to cancer clusters in various workplaces around the country.

Potential remedies: See chapter 10.

Have you tested the AC magnetic fields and other EMFs (such as electric, RF, and dirty electricity)? Are the AC magnetic fields higher than 1 mG? (2)

☐ Yes ☐ No ☐ Don't Know

What the dangers are: Any reading at 1 mG and above can trigger small but in some cases seriously detrimental changes in your body. The *BioInitiative Report* maintains that AC magnetic fields should not exceed 1 mG. Compare this with the AC magnetic fields found in nature, which are less than .0000002 mG!

Potential remedies: See chapter 10.

Have you used a cell phone more than four hours a day for ten years or more? (3)

☐ Yes   ☐ No   ☐ Don't Know

What the dangers are: Studies from Sweden, where there's a long history of cell phone use, have found brain cancer risks rise after ten years of "heavy use." Scientists are now concerned that with cell phone use starting in childhood the rates of brain tumors may rise even higher in younger people.

Potential remedies: See chapter 10.

Are there power lines or transformers within one hundred feet of where you work? (2)

☐ Yes   ☐ No   ☐ Don't Know

What the dangers are: Power lines and transformers, which have been linked to leukemia, create a large EMF that dissipates with distance.

Potential remedies: See chapter 10.

Are you an electrical line worker, an electrician, a welder, or do you work in a radiology lab? (2)

☐ Yes   ☐ No   ☐ Don't Know

What the dangers are: Studies have found that people who work in these professions have higher rates of serious diseases than the rest of the population.

Potential remedies: See chapter 10.

Do you work with your laptop on your lap? (1)

☐ Yes   ☐ No   ☐ Don't Know

What the dangers are: While your laptop is connected to the AC power adaptor, it can cause high AC magnetic and AC electric field exposure. When your computer is on your lap, that exposure is right next to reproductive and other organs.

Potential remedies: See chapter 5.

Guidelines and Standards (Worldwide) for AC Magnetic Fields

From electrical engineer Larry Gust:

1. ACGIH[1] occupational TLV[2]: 2000 milliGauss (mG)

2. DIN[3]/VDE: occupational 5000 mG, general 4000 mG

3. ICNIRP[4]: 1000 mG

4. Switzerland: 10 mG

5. WHO[5]: 3–4 mG "possibly carcinogenic"

6. TCO[6]: 2 mG

7. U.S.-Congress/EPA: 2 mG

8. BioInitiative Report[7]: 1 mG

9. As found in nature: <0.000002 mG

----

1. American Congress of Governmental and Industrial Hygienists
2. Threshold Limit Value
3. German Standards Institute
4. International Commission on Non-Ionizing Radiation Protection
5. World Health Organization
6. Swedish Confederation of Professional Employees
7. www.bioinitiative.org

The original chart, from "EMF in Your Environment," EPA, 1992, contained measurements for video display terminals that are obsolete technology, so I didn't include them here.

Do you work within one hundred miles of a nuclear power plant? (1)

☐ Yes   ☐ No   ☐ Don't Know

What the dangers are: You may be exposed to very low levels of radioactivity.

Potential remedies: See chapter 11.

Total: _____

At School

Is your/your child's school located near cell phone or radio frequency towers? (2)

☐ Yes   ☐ No   ☐ Don't Know

What the dangers are: Studies have found increased risk of diseases and tumors in people living within about twelve hundred feet of these towers. Your children may be spending six or more hours a day bathed in low level radiation.

Potential remedies: See chapter 9.

Is your/your child's school set up for Wi-Fi? (2)

☐ Yes   ☐ No   ☐ Don't Know

What the dangers are: There's no escaping the harmful waves produced by wireless broadcast systems. If your home is also wireless, your child may be exposed 24/7.

Potential remedies: See chapter 9.

Is your/your child's classroom adjacent to the school's computer laboratory? (1)

☐ Yes   ☐ No   ☐ Don't Know

What the dangers are: Modern computers don't produce as much electromagnetic radiation as older models, but a concentration of devices in one room can create significant fields of various frequencies which can penetrate walls, especially when each computer is operating with wireless Internet connections!

Potential remedies: See chapters 9 and 10.

Total: _____

At Outdoor Play and Recreation

Are there power lines within one hundred feet of where you play or go for outdoor recreation? (2)

☐ Yes   ☐ No   ☐ Don't Know

What the dangers are: Electrical equipment can produce significant fields within one hundred feet, though they dissipate the farther you move away.

Potential remedies: See chapter 6.

Are there transformers within fifteen feet of where you play or go for outdoor recreation? (2)

☐ Yes   ☐ No   ☐ Don't Know

Potential remedies: See chapter 6.

Do you play or recreate within one hundred miles of a nuclear power plant? (1)

☐ Yes   ☐ No   ☐ Don't Know

What the dangers are: You may be at risk of exposure to very low levels of radiation which may leak from nuclear plants—even more so if you're near a radioactive waste dump site.

Potential remedies: See chapter 11.

Total: _____

SCORING THE QUIZ

If your score is 15–25, you have a high risk of being zapped.

If your score is 6–10, you have a moderate risk of being zapped.

If your score is 5 or less, you have a lower risk of being zapped.

If you are feeling overwhelmed, you're definitely not alone. Most of us have never questioned the health implications of all our modern-day technology and conveniences. We just take it all for granted. We buy what industry advertises, what others have, or what sounds hot, not realizing that some of these products may be zapping us of our vitality. Hopefully, with some guidance throughout the next chapters, you will have the facts you need to make more informed choices, to help you and everyone you care about.

ZAP-PROOF YOUR HOME

The quiz you just took in chapter 4 has given you some idea of where you've been zapped. The good news is, there are a number of practical EMF solutions and suggestions that are relatively easy and inexpensive. By all means, take advantage of them to create a healing haven in your home. I like to call this process electronic feng-shui because it can involve moving furniture and appliances to help establish a lower EMF environment.

You can choose to do as little or as much as you want, but I would recommend that you focus your energy on the rooms where you spend most of your time: your bedroom, your living room, your family room, and your home office.

Here's what you need to do, room-by-room.

YOUR BEDROOM

This is the room where you should be spending at least eight hours sleeping at night. If you're tossing and turning instead, chances are, you're being kept awake by the unnatural, man-made voltage in your body caused by electrical fields coming from the wiring in your walls, the ex-

Get the Lighting Right

Thomas Edison's incandescent light bulb was a good idea, and though there have been subsequent generations of bulbs, it's still the best idea, say most EMF experts.

Incandescent lights work fairly simply: when 120 V of electricity hit a metal (tungsten) filament, it lights up. Fluorescent lights work when electrons strike mercury vapors in the tube, causing them to emit ultraviolet light. When they hit the phosphor coating on the tube, it produces a white light. They're also equipped with a device called a ballast, or transformer, which contains electronics that emit a much higher magnetic field than the simple incandescent bulb—6 mG at two inches away compared to 0.3 mG at two inches. These fluorescent lights also emit a strong electric field at multiple frequencies from 60 Hz into the low RF range. A portion of these emissions includes dirty electricity. Which one do you want next to your head?

Halogen light bulbs work similarly to incandescents, though the use of halogen gas inside a quartz (not glass) "envelope" allows them to burn longer because they don't wear away the tungsten filament as quickly—in fact, the halogen re-

tension cords or power strip under your bed or desk, or the cords from your bedside lamps, clock, and other appliances. Since your body restores itself overnight—that's also when your body produces 80–90 percent of its melatonin—you need to be scrupulous about reducing the sources of EMF in this room even more than the others in your house.

Clean up your bedroom. Not the clutter, the electronics. Since the greatest healing occurs during sleep and you spend nearly one-third of your life in your bed, the bedroom is the most important room of the house to zap-proof. That includes TVs, radios, clock radios, alarm clocks (except the battery-operated kind), cordless phones, mobile phones, heating pads, and older electric blankets. They need to be *out* of the bed-

turns the tungsten to the filament instead of vaporizing it away as incandescents do. That makes halogens far more economical. But they have several problems. First, they generate a lot of heat and need good ventilation so they don't become a fire hazard. Lamps generally have a transformer in their base, which generates high AC magnetic fields. If you use them in recessed lighting, the field isn't as much of an issue underneath as it is above, depending on the location of the transformer, so make sure you position beds and other furniture where you spend a lot of time at least three feet away from the location of the transformer. And make sure you turn them off when not in use.

The new green, energy-efficient bulbs are compact fluorescents that come with many health hazards—mercury, UV radiation, RF fields, and dirty electricity. They're referred to as intermediate frequency by the World Health Organization, and these "middling" fields are biologically active. A *New York Times* story details quality control problems with the bulbs—possibly caused by the rush to bring them to market, cheap parts, and poor installation instructions.[1]

Better yet consider LED bulbs. They are mercury free and RF free, without the dangers of dirty electricity. The Geode Bulb, a form of LED, is currently becoming more widely available, so keep an eye on LED light development.

room or at least as far from you as possible. Some people even turn off the electric power to their bedrooms at night.

Move the bed. If you can't cut power to your bedroom at night (if your smoke or carbon monoxide detector is hardwired to the circuit, you don't want to do that), make sure your bed is positioned so that your head isn't near a power outlet and be aware of any AC magnetic fields that might emanate from below or next to you, like those from a fluorescent or halogen light fixture, power mains, refrigerator, computer, or other appliances. If it's possible and your room is large enough, move the bed away from the wall because that's where the electrical wiring of your house lives. You want to keep your body as far away (a minimum of three feet) from the fields as possible while you're sleeping. Since metal transmits electric fields, you may also want to consider replacing your metal spring mattress or head-board with something made of natural material, such as a futon. If your bedroom is on a second floor, there may be no escaping wiring—it's in the floor. In that case, your best, least expensive option for protection is to cut power to any wiring that is above, below, or around your bedroom during the hours you are asleep.

Use a battery-operated alarm clock. If you can't keep up with the battery-changing, make sure your electric clock or clock-radio is at least six feet from your head—meaning you'll actually have to get up to turn it off, which might be good in many ways.

Move the baby monitor. Baby monitors also emit high magnetic fields, but if you've reduced your (and your baby's) exposure in other ways, it may be worth keeping—if just for your peace of mind. Make sure to keep it at least six feet from crib and bed. Wired, of course, is preferable to wireless.

Get earthed. The illustrious American-sponsored Tour de France Cycling Team used mattress pads woven with silver threads that are grounded to the earth, much like your electrical system, with a steel rod plunged into the ground. The team chiropractor, Jeffrey Spencer, reports that the team slept better and had more energy, less tension, more strength and stam-ina, and healed faster when they were earthed.[2] See "Connect Yourself to Mother Earth" on page 69 to learn about Earthing technology.

Turn off the electric blanket. Early studies by Nancy Wertheimer and Ed Leeper found that electric blankets had an electrical field of 5 mG or more that penetrates six to seven inches into the body.[3] Your best bet is to use it to warm up your bed before you get in, then turn it off and unplug it (it still generates a field when it's plugged in). Waterbeds, like electric blankets, are in very close proximity to the body, which can also negatively impact it with electrical fields.

Check your lighting. Energy efficient fluorescent lighting, both the in-the-ceiling long tube type or the compact fluorescent (CFL) variety (with the corkscrew tube), produce elevated magnetic and electric fields as well as dirty electricity when they are on. If you happen to break one bulb, you can have a Hazmat crisis: the mercury in one bulb is enough to contaminate the entire room, says Larry Gust, an electrical engineer and certified building biologist whose company, Gust Environmental in

California, does indoor environmental inspections and mitigation. "And what about the environmental effect of sending millions of spent bulbs to landfill?" he notes. Avoid these bulbs and use old-fashioned incandescent bulbs until a wider variety of LED lights are available.

Reduce exposure to fluorescents, halogen torch lights, and even CFLs—they all generate significant EMFs. Most standard lamps give off a high electrical field regardless of how many light bulbs are used or their wattage, so position them as far away from your head as you can when you're sleeping. Either unplug the lamp or install a timing device that turns the power off during sleep. A low-voltage DC halogen light that's set into the ceiling is safe if you're beneath it, but it can radiate high magnetic fields about three feet into the floor above, so don't place them underneath rooms where children or adults sleep or play.

Rethink those fans. Ceiling fans operate with an electric motor that will generate high magnetic fields in the room above, so they're not appropriate in rooms beneath a child's bedroom or playroom. Free-standing electric fans are fine as long as you keep them at least three feet from your body. So are fan, convection, and oil-filled radiators as long as you keep them at the same distance. Remember the rule of thumb: location, location, location. Exposure is reduced by 85 percent every time you double the distance from the source of the magnetic field.

Invest in room-darkening drapes. They won't reduce your exposure to electrical and magnetic field radiation, but by keeping your room darker longer, they can help your body produce more melatonin, the antioxidant hormone that is diminished by exposure to light and to EMFs.

YOUR KITCHEN

Though your kitchen contains more electric devices per square foot than any other room of your home, most of them—like your toaster, coffeemaker, or blender—produce no fields unless they're in use. Even if you like both your toast and coffee really hot and black, the appliances are just not on long enough to do you harm.

But your stove, refrigerator, microwave, and dishwasher contain electronics that require a transformer—and that runs as long as the appliance is plugged in. When they're turned on, they produce extra strong AC magnetic fields, so you need to give them a wider berth. Fortunately, even if you're a budding Julia Child, you won't be spending that much time in the kitchen and certainly not hanging around the refrigerator. Just keep small children out of the kitchen when everything's going and pay attention to the rooms beneath and on the other side of those appliances—the magnetic fields they create extend through walls and floors.

Inspect your microwave oven. When you say you're going to nuke dinner, everyone will know you're heating food in the microwave. Microwave ovens don't use nuclear power, but they do use RF power like radio and TV broadcast towers, your cell phone, and radar. (You've heard the term *microwave tower* when referring to cell towers—they're all using the same high frequency energy as the oven where you heat up your Lean Cuisines.) Some experts think they're more dangerous than they should be. Like Larry Gust.

"Unfortunately," Gust says, "the microwave oven is designed to meet government standards on RF leakage. These standards are thousands of times too high. So these units leak radiation right through the window in the door and this energy travels great distances in free air. This RF leakage is avoidable by where you position the microwave. But RF radiation is not the end of the story. The nutritive value of microwaved food has been destroyed by the violent vibration of the water molecules (2.4 billion Hz) in the food. I am sorry to say this, but you are better off not eating than to eat microwaved food."

As your microwave gets more use, the door can become loose, the seals can wear out, and the safety interlock devices that keep the microwaves from generating while the door is closed can stop working correctly. This is not a DIY job. You need to get a qualified service person to inspect and repair your microwave (or, better yet, replace the old one, if you decide to have one at all, since the cost is minimal). No matter what, don't stand in front of the microwave and watch your tea water boil or your popcorn pop.

If your microwave is positioned conveniently over your cooktop, remember that it's also around head level. You should locate it elsewhere; if that's not possible, don't use it while you're working at the stove.

Connect Yourself to Mother Earth

As a pioneer in the U.S. cable TV industry, Clint Ober was intimately familiar with the role of grounding to protect the integrity and clarity of transmissions. He knew the problem that electromagnetic fields could cause. To prevent signals from escaping and picture-distorting noise from penetrating, cable systems had to be shielded and then grounded—connected to the earth—to prevent interference.

After his retirement, Ober began to wonder if there was any need for humans, like all electrical systems—including cable TV—to be grounded in order to function effectively. He had become aware that most humans wear shoes with plastic or rubber soles that insulated them from the earth's natural electrical field, the same field that provides the stable ground for electrical systems. He wondered if human health could be affected in some way by this separation. Plastic and rubber cannot conduct electrical energy. Humans evolved barefoot, he thought. They walked barefoot, or used hides (leather) on their feet, permitting conduction of earth's surface energy into their bodies. They slept on the earth, or on hides. It would be likely that nature would "have used that electrical resource in some way," he thought.

Ober's curiosity motivated him to explore whether we somehow needed the connection to our mother earth in order to be healthy. He started grounding himself and checking his own electrical energy with a simple voltmeter at various spots in his house. He noted how his body voltage changed—high when exposed to EMFs, at his computer, and in his bedroom, but at practically zero voltage when he grounded himself by holding a wire connected to a ground rod placed in the earth outside. He then devised a crude conducting grid that he placed on his bed, dropped a grounding wire attached to it out the window, and secured it to the earth via the outside rod. He lay down, voltmeter in hand, and fell asleep! He was in the same position when he woke up the next morning, the voltmeter still on his chest. He had had his first sound sleep in years and hadn't needed to resort to painkillers, which was often the only way he could fall asleep. He repeated his experiment and within a week or so noticed that the pain and stiffness from previous skiing injuries and back surgery were gone. *(continued)*

Astonished, Ober tried the same experiment on friends who had sleep and pain problems. They had the same positive experience. Ober then set out to find doctors and researchers who knew something or could explain what he had found. Nobody was interested. Nobody knew anything. So he decided to use his own funds to find the answers. Was it for real? How did it work for people?

It's now been more than ten years since he began pursuing scientific validation, first on his own and then with established researchers. He's still at it. The evidence uncovered to date is extraordinary. Personal grounding, or Earthing as he calls it, promotes better sleep, more energy, quicker healing, and reduced inflammation and pain, and normalizes production of the stress hormone cortisol.[4]

A dozen studies confirm the theory that people, like cable TV and all electrical systems, need to be grounded. The findings support the pioneering work of Dr. Ross Adey, who believed that EMFs interfered with the natural, normal electrical communications between cells. Personal grounding prevents the interference of outside noise in normal cellular "whispering," which can lead to the kinds of signaling errors that can cause cancer and damage the immune system. The research demonstrates that Earthing maintains the human body at the same voltage of the earth. The body receives a stabilizing electrical influence for all its many bioelectrical circuits. We also receive an infusion of negative charged free electrons, omnipresent on the surface of the earth as a result of solar radiation and lightning strikes. They are transferred into the body when we make physical contact with the earth and appear to neutralize positively charged free radicals that cause inflammatory damage at the cellular level.

In the Earthing studies, Ober and a number of researchers—at least one of them a skeptic—found that grounding:

- Reduced nighttime levels of cortisol. Maurice Ghaly, M.D., and Dale Teplitz, M.A., led an investigation of twelve people with chronic pain and poor sleep and found that sleeping earthed reduced pain and stress, and promoted sleep.[5] Cortisol, which is released during emotional or physical stress, was brought into a more normal twenty-four-hour circadian rhythm. "Cortisol is one of the reasons for poor sleep," says Ober. "Worry and anxiety stimulate cortisol secretion [which] interferes with sleep and you wake up feeling like you have acid in your veins."

The study, and others that followed, also provided a new clue to Earthing's ability to reduce pain: cortisol is also linked to inflammation. In recent years, scientists have begun to think of inflammation as the unidentified link in diseases as varied as arthritis, cancer, heart disease, and Alzheimer's. Your body's immune system mobilizes white blood cells, an array of chemicals, and an initial inflammatory response to protect against infection and foreign invaders. This response, however, can turn from friend to enemy if it continues unchecked. Then, it may cause painful arthritis, asthma, colitis, atherosclerosis, and even cancer. Inflammation plays a role in chronic pain syndromes.

Interestingly, Ghaly, an anesthesiologist in California, undertook the study because he *didn't* believe it would work. His surprising (to him) findings were published in 2004 in *The Journal of Alternative and Complementary Medicine.*

- Promoted—quickly—greater peripheral blood flow, relaxation of the muscles, and calming of the left hemisphere of the brain (typically thought of as the thinking side). An experiment by Gaétan Chevalier, Ph.D., then director of research at the California Institute for Human Science in Encinitas, earthed fifty-eight study subjects and used encephalography (which measures brain waves) and other sophisticated measurements to monitor the effect of Earthing on a variety of body functions aside from cortisol synchronization, and pain and stress reduction. The immediate response by the left brain as well as in the blood flow indicate a general relaxation, which may explain why grounding seems to promote better sleep.[6]

Beginning in 2000, Ober began developing a variety of products, including mattress, floor, and desk pads as well as travel items that bring the earth's natural energy inside for use in homes and offices. I've been using them for more than a year, and I've never slept better (see resources).

During his research, Ober learned that the steep rise in inflammation-related diseases started after the 1960s when we started wearing rubber or plastic-soled shoes (replacing conductive leather) and installing wall-to-wall carpeting in our homes. "That," he says, "was when the big disconnect occurred."

Ober is quick to point out that anybody can ground oneself anywhere and connect to Mother Earth. "All you have to do is take your shoes off and put your feet on the earth," he says with a laugh. Simple, but good advice. Go barefoot as often as possible or sit with your bare feet on the ground—it's what your Mother would like you to do.

Don't linger at the fridge. Inside your refrigerator there's a big electric motor that runs the miniature air conditioning system that keeps the food cool. You can hear that motor when it starts. When it stops, you'll notice that the kitchen is suddenly a lot quieter. This motor uses more power than lights and electronics, says Gust. Additionally, there may be a fan that circulates air inside, and it can produce a magnetic field of several milliGauss near the refrigerator. There may even be an electric door heater to prevent condensation. The fields they create are usually not a big problem. "People don't usually spend a lot of time leaning on the fridge," says Charles Keen, EMF mitigation specialist. But if your sofa is sitting on the other side of the wall, you and whoever sits on it will be bathed in a massive field when the refrigerator is running. If you can't move the fridge, move the sofa. If you're planning a kitchen, keep the refrigerator at least three feet from cooking or eating areas and away from the wall with the sofa in the family room.

Studies of these so-called intermittent EMFs are troubling. Charles Graham, an experimental physiologist at the Midwest Research Institute in Missouri, and his colleagues found that the kinds of on-off fields we're exposed to from refrigerators and dimmer switches disturb sleep and sleep quality, and decrease heart rate variability (HRV). (HRV is a hallmark of a healthy heart: your heart rate needs to respond to your central nervous system, so a steady heart rate can reflect that something's gone wrong.) In Graham's studies, participants slept well and had normal HRV while exposed to background or constant EMFs. Not so when they were sleeping in a room bombarded by intermittent waves.[7] See chapter 7, "The Newest Zapper: Dirty Electricity."

Cook at arm's length. If you are using a gas stove, it's not producing electrical and magnetic fields, even if it has an electronic switch. If, however, you use electric, you should stay at least three feet from the oven while it's in use, but that's often not practical, especially if you're sautéing on the stove top. In that case, try to use back burners as much as possible so you can at least be a little farther away. Also, be aware that the range hood, which usually contains a fan and a light, can also emit a good-size field up to about a foot and a half away, and it's usually at head height. So only use the fan and light when it is absolutely necessary.

Clean at a distance. Self-cleaning ovens save you hours of messy scrubbing, but because they use high wattage to literally burn debris off oven walls, they also emit high AC magnetic fields. While your oven is cleaning, stay out of the kitchen.

Survey your gadgets. Coffeemakers, mixers, toasters, breadmakers, electric can openers, countertop grills, and electric teakettles can all give off EMFs, but your exposure to them is likely to be quite brief and you can be protected by the distance you are from the device. However, you may want to do things the old-fashioned way with substitutes that are not electrically powered.

YOUR BATHROOM

It may be the smallest room in your house, but it holds some of the biggest EMF producers you own—some of which may surprise you. But, except for those long soaks you take, your exposure is very, very low.

Replace your hair dryer. It may be small, just like your flatiron and curling iron, but all of these styling tools pack a powerful EMF wallop to the tune of a 25 mG field around the motor and even more around your hand and head. The reality is that EMFs can be biologically harmful at much lower levels like 1 mG and below.

There are a number of new hair dryers available that clock in at around 2 mG. One I like in particular is the Chi Ceramic Low EMF Ionic Hair Dryer with a six-foot cord that will keep you well away from the 450 mG motor to make your actual exposure only about 0.1 mG for as long as it takes you to dry and style your "do." It's a good choice for professional hairstylists who have far more exposure than the average person.

Go manual with the razor. When the Electric Power Research Institute (EPRI) measured the magnetic fields around common household electrical devices, one of the eye-opening findings was the electric razor. Granted, it produced a very large field a mere six inches away. Since it's used at *no* inches away, the real-world field is likely to be even higher. But, like the hair dryer, it's not an appliance we spend hours with, so the actual impact on our health may be less because there is reduced exposure. Still, you can always err on the side of caution and go back to manual razors, of course.

YOUR LIVING/FAMILY ROOM

Like your kitchen, your living or family room might be appliance central, with TVs, audio and gaming systems, overhead fans, maybe even an electric fireplace. But you don't have to give it all away. Most of the solution in these rooms involves just a little heavy lifting.

Move away from the wall. It's tempting to arrange furniture around the perimeter of these rooms, but if you and your family are going to spend lots of time in them, pull seating away from the walls. Your electrical wiring is located there, and you may be exposed to a source of high EMF such as a heater motor, electric hot water heater, refrigerator, or freezer.

Buy a new set. If you still have a bulky conventional TV set, you're exposing yourself to more EMFs than those who have traded up to models with LCD screens, which have much lower fields and better pictures.

Back away from the TV. When you were a child, your parents may have cautioned you to sit six feet from the TV to protect your vision. They probably read it in the owner's manual, where the advice really had nothing to do with preventing you from needing glasses. It was just a suggestion to optimize your viewing pleasure—the picture actually looked better from far away because the lines that made it up were then invisible. Unwittingly, your parents protected you from the large EMFs surrounding the pre-modern sets. Newer sets have lower EMFs, but it's still a good idea to keep a distance—six feet is good. Fortunately, depending on what size screen you have, that's still recommended to improve picture quality.

Stay away from wires. Power cords and surge protectors can also project electric fields. Surge protectors contain several transformers that can create a magnetic field extending two to three feet away. Gaming systems have a transformer that plugs into a power strip or socket and gives off very high AC magnetic fields. Even the controllers can give off a high RF field—check gaming equipment with your TriField or Gauss meter (see resources).

Invest in power strips. Even if your electronics are off, they're still generating an AC magnetic field if they're plugged in. An electronic equipment or electronics-containing appliance (does it have a digital display?) doesn't ever turn off—it just goes into standby. That includes satellite receivers for your TV or VCR, answering machine, microwave oven, and more. They could be draining your electricity when they're not in use (costing, according to one estimate, 10 percent of your electric bill).[8] A power strip will allow you to shut them all off at once. Also, new power strips by Belkin, available at hardware stores, will factor out high frequencies between 150 kHz and 100 MHz.

YOUR HOME OFFICE OR DEN

Chances are, between your computer, mouse, keyboard and printer (wireless Internet and ELF issues), and cordless phone, you're regularly surfing significant RF waves.

Get wired. Consider switching back to a wired network. Yes, that means adding back a few wires to that snake of cables already under your desk and giving up a little convenience since you'll have to wire all the computers in your house, including your laptop. You won't be able to roam the house, laptop in hand. But in a wireless system, you need a router that is basically a radio device that receives and sends signals to all the computers and other technological devices—like your printer—on your system. The point is, you need to buy a wired router and run Internet cords to as many workstations in your home as needed.

Go back to corded phones. If you have a cordless phone and it's using DECT technology, it's constantly transmitting RF radiation even when you're not using it. (Check your phone—if it's 2.4 or 5.8 GHz, it uses pulsed digital signals, which are more damaging than earlier analog signals.) Switch to a conventional corded push-button phone. They are still available—many businesses use them—for the majority of your talk time. Cordless phones emit a very significant amount of radiation. "Why put all that RF power through your head when you don't have to?" says Gust. "What is the price you are willing to pay down the road to have a little bit of extra convenience right now?"

Don't cradle your laptop. It may be called a laptop, but don't use it in your lap at any time. It radiates harmful EMFs whether it's connected to the AC power adaptor or not. Better yet, put it on a table or desk. You can add a helpful choice. Get a shielding pad to place your laptop on when you set it on your lap (see resources). Turn off the wireless connection in your computer software when it is not being used.

Choose the right computer monitor. If you still have an old-style monitor (called a cathode ray tube), buy an LCD monitor, which has comparatively reduced EMFs. Most of the information you'll read about the danger of computer monitors refers to cathode ray tubes, which aren't used much anymore. If you have an old computer, consider purchasing a new screen.

Adjust your chair. And not just for good ergonomics. You want to position the chair you use for working at the computer (where you're likely to spend most of your time) away from the wall where wiring or another source of EMF is located, such as the base unit of a desktop computer. Keep wiring, surge protectors, power bars or extension cords, and ideally the desktop computer tower, about five feet away from you, and position any fax machines or other office equipment as far away as possible (three feet minimum would be best).

EPA MAGNETIC FIELDS MEASUREMENTS CHART

Magnetic field measurements are in units of milliGauss (mG)

	Distance from Source			
	6 inches	1 foot	2 feet	4 feet
BATHROOM				
Hair Dryers				
Lowest	1	—	—	—
Median	300	1	—	—
Highest	700	70	10	1
Electric Shavers				
Lowest	4	—	—	—
Median	100	20	—	—
Highest	600	100	10	1
KITCHEN				
Blenders				
Lowest	30	5	—	—
Median	70	10	2	—
Highest	100	20	3	—
Can Openers				
Lowest	500	40	3	—
Median	600	150	20	2
Highest	1500	300	30	4
Coffeemakers				
Lowest	4	—	—	—
Median	7	—	—	—
Highest	10	1	—	—

	Distance from Source			
	6 inches	1 foot	2 feet	4 feet
KITCHEN *(continued)*				
Electric Slow Cookers				
Lowest	3	—	—	—
Median	6	—	—	—
Highest	9	1	—	—
Dishwashers				
Lowest	10	6	2	—
Median	20	10	4	—
Highest	100	30	7	1
Food Processors				
Lowest	20	5	—	—
Median	30	6	2	—
Highest	130	20	3	—
Garbage Disposals				
Lowest	60	8	1	—
Median	80	10	2	—
Highest	100	20	3	—
Microwave Ovens				
Lowest	100	1	1	—
Median	200	4	10	2
Highest	300	200	30	20
Mixers				
Lowest	30	5	—	—
Median	100	10	1	—
Highest	600	100	10	—

| | Distance from Source | | | |
	6 inches	1 foot	2 feet	4 feet
KITCHEN *(continued)*				
Electric Ovens				
Lowest	4	1	—	—
Median	9	4	—	—
Highest	20	5	1	—
Electric Ranges				
Lowest	20	—	—	—
Median	30	8	2	—
Highest	200	30	9	6
Refrigerators				
Lowest	—	—	—	—
Median	2	2	1	—
Highest	40	20	10	10
Toasters				
Lowest	5	—	—	—
Median	10	3	—	—
Highest	20	7	—	—
LIVING ROOM/ FAMILY ROOM				
Ceiling Fans				
Lowest	—	—	—	—
Median	—	3	—	—
Highest	—	50	6	1
Window Air Conditioners				
Lowest	—	—	—	—
Median	—	3	1	—
Highest	—	20	6	1

| | Distance from Source | | | |
	6 inches	1 foot	2 feet	4 feet
LIVING ROOM/ FAMILY ROOM (continued)				
Turntables/Tape Players				
Lowest	—	—	—	—
Median	—	3	1	—
Highest	3	1	—	—
Color TVs				
Lowest	—	—	—	—
Median	—	7	2	—
Highest	—	20	8	4
LAUNDRY/UTILITY ROOM				
Electric Clothes Dryers				
Lowest	2	—	—	—
Median	3	2	—	—
Highest	10	3	—	—
Washing Machines				
Lowest	4	1	—	—
Median	20	7	1	—
Highest	100	30	6	—
Irons				
Lowest	6	1	—	—
Median	8	1	—	—
Highest	20	3	—	—
Portable Heaters				
Lowest	5	1	—	—
Median	100	20	4	—
Highest	150	40	8	1

	Distance from Source			
	6 inches	1 foot	2 feet	4 feet
LAUNDRY/UTILITY ROOM *(continued)*				
Vacuum Cleaners				
Lowest	100	20	4	—
Median	300	60	10	1
Highest	700	200	50	10
BEDROOM				
Digital Clock				
Lowest	—	—	—	—
Median	—	1	—	—
Highest	—	8	2	1
Analog Clock				
Lowest	—	1	—	—
Median	—	15	2	—
Highest	—	30	5	3
Baby Monitor				
Lowest	4	—	—	—
Median	6	1	—	—
Highest	15	2	—	—
WORKSHOP				
Battery Chargers				
Lowest	3	2	—	—
Median	30	3	—	—
Highest	50	4	—	—

| | Distance from Source | | | |
	6 inches	1 foot	2 feet	4 feet
WORKSHOP *(continued)*				
Drills				
Lowest	100	20	3	—
Median	150	30	4	—
Highest	200	40	6	—
Power Saws				
Lowest	50	1	—	—
Median	200	40	5	1
Highest	1000	300	40	4
Electric Screwdrivers (while charging)				
Lowest	—	—	—	—
Median	—	—	—	—
Highest	—	—	—	—

Source: "EMF in Your Environment," Environmental Protection Agency (EPA), 1992.

ADVANCED ZAP-PROOFING

Now that you have started your EMF cleanup on the homefront, you may find yourself really motivated to identify the "hottest" spots inside and outside your home to measurably reduce your toxic load. We'll continue to reference the quiz from chapter 4 as we explore reducing EMF and RF exposure even further.

While awareness is absolutely paramount to change, actually measuring your true exposure will make EMF remediation that much more specific and effective. For this, you will need some tools.

BE YOUR OWN GHOSTBUSTER

There is a dizzying array of technological devices to help you detect and measure EMFs, and you might be surprised to find yourself shopping at the same places that electricians, health and safety experts, and paranormal investigators buy their gear. (Yes, your friendly local Ghostbuster uses an EMF detector to suss out spirits, though how they distinguish otherworldly visitors from the EMF smog in some American households, with their twenty-some appliances, computers, cordless phones, and proximity to cell and broadcast towers, I don't know.)

The least expensive option is an ordinary, battery-operated AM transistor radio, which will pick up RF fields (though not the typical 60 Hz fields). Simply tune it between stations and walk around your house, listening for the various changes in the static that will indicate electrical fields, from which you can assume there are also magnetic fields. I wouldn't go out and buy one—it's not the best meter—but if you already have a transistor radio, it can be your eye-opening introduction to the invisible pollution in your home or office. An AM radio will pick up the RF fields from 55 to 1600 kHz. When it does, the static sound will get louder. Some areas—like an office with a computer and wireless router—will produce stronger static sounds. Using the radio, you can even judge at what distance the fields begin to grow weaker (the static will start to fade), which can help you determine where to position sofas and chairs in relation to the television set to minimize exposure.

When Larry Gust tested an AM radio at his house, he says, it detected "a 2.4 GHz router and a cell phone charger. Other things—a copper coil transformer (60 Hz), a 900 MHz cordless phone, high magnetic fields from my decorative lighting transformer (60 Hz) on the other side of the wall barely registered, if at all."

To pick up even more RFs without spending lots of money, Gust recommends a telephone amplifier made for the hearing impaired, available at RadioShack for as little as ten dollars.

If you want more precise measurements of 60 Hz fields, you can purchase a Gauss meter (at around two hundred dollars for a good basic model). This device, which is small enough to carry in your pocket when you go appliance or computer shopping, measures the strength of a magnetic field, which is expressed as gauss or tesla. Standard guidelines for safe exposure are no more than 2 to 3 mG, though studies have found biological effects at far below that, particularly in children. Use the Gauss meter to measure the field around your electronics and appliances until it reads 1 mG or less, which is what most EMF experts now believe is relatively safe. That gives you a good indication of the distance you and your family should be away from your TV, computer monitor, and even your humidifier. You can also use it to measure magnetic fields around power lines—or, if you're house-hunting, to pick up the ambient EMFs in and around a potential dream home. (To tell you the truth, there are dozens of

Gauss meters available on the market at various price points and with an assortment of bells and whistles. You'll see them described as either single- or triple-axis coil. The single-axis models are cheaper, but you have to point the meter's sensor in three directions around an EMF, and then combine the three readings to give you the combined field strength. A triple-axis meter requires only one reading. There are also specialized meters that, for example, pick up just RFs.)

According to Gust, a flat-response Trifield meter that, as its name suggests, detects and measures electrical, magnetic, and RF fields, costs around one hundred thirty dollars. It's great for detecting magnetic fields, but it isn't quite as sensitive on electrical fields and is considered unreliable for RFs.

All of these products are available online. Before you make your purchase, discuss with a knowledgeable salesperson or customer service representative what you want to measure and what you plan to do with the information you get from the device so he or she will be able to direct you to the right product for you. Remember—expensive isn't always better.

When I asked Gust to recommend a home meter for me to use, he suggested a triple-axis AC magnetic field meter from www.lessemf.com, a leading supplier of EMF detection and mitigation products. The 60 Hz Blocking EMF Meter has an LED screen that's easy to read and is calibrated to range from 0.01 to 40 mG—and costs less than one hundred seventy dollars. The most expensive devices are those designed for industrial use, which would be overkill in a private home.

If you are using an EMF consultant, it is important to ascertain their experience in assessing all key fields: electric, magnetic, RF and microwave frequency, and dirty electricity.

WHAT TO MEASURE

While studies have found that most homes are below 3 mG and many are below 1 mG, those are average numbers. Your home may not be average. You may have localized fields that are much higher—as high as 10, 20, even 50 mG is "quite common," says physicist and EMF expert Ed Leeper in his book *Silencing the Fields*.[1]

Leeper cautions against making the mistake that some field researchers do: measuring in the middle of the room in the middle of the day. That's not going to tell you much. Most of us are exposed to EMFs around the perimeter of our rooms where we place our TVs, computers, appliances, and where the wiring is located. And all of those things are more likely to be humming at night, when everyone is home and all our lights and appliances are on, than they are during the day.[2]

YOUR OUTSIDE WORLD

But before you start inside the house, go outside. Look around. Do you see any transformers? They're either the barrel-looking devices at the top of poles in front of or behind your home, or a metal box on the ground if the wires are underground. Look for larger pad-mounted metal ground transformers painted with the words *High Voltage*. Keep in mind, though, in the studies linking increased rates of leukemia to proximity to electrical transformers, the electrical and magnetic fields were coming from the wires, not the box itself.

What about transmission lines? Those are the power conduits carried on pylons throughout the country bringing energy to homes and businesses from the source. The bane of real estate agents everywhere, these massive electrical towers usually scare away most house hunters except for those who are willing to sacrifice health and home appreciation for a bargain-basement selling price. The EMFs they produce are so powerful that British artist Richard Box installed 831 fluorescent tubes under power lines in Bristol, England—just stuck them in the ground—and they lit up like birthday candles. He called his artwork "Field."[3] If you stand directly under power lines, you may feel a mild shock if you touch something that conducts electricity—and that includes another person. The closer you live to transmission lines, the greater your exposure to their electrical and magnetic fields, though the fields they generate can vary dramatically in strength in a day. They weaken the farther away you move.

You might be inclined to think that your yard is free of EMFs, but there's a little-known (to homeowners, at least) source of these fields on your home's water line.

But before we get to your plumbing, a little electrical lesson first. The power to your home arrives from its source on one line in the form of current flow (your electrician measures this in amps), and the spent electricity returns to its source on a second right next to it. A single wire carrying a power source of even one amp can create a powerful magnetic field that spreads out into a large area. A second wire that carries an equal amount of current flowing close to it in the opposite direction will mostly cancel that field. There will still be a small magnetic field, but it will be localized to the wire. That's why you need two wires—an in and out, if you will—and they have to carry the same size current, or the resulting imbalance will create a large magnetic field, called a net current.

That's why all your appliances, from lamps to fridge to hot tub, have a cable or cord that contains two wires inside—one to bring you power, one to return it. That return wire is called the neutral. So what does this have to do with your plumbing? You probably know that your electrical system neutral wire is connected to the metallic water piping in your home (part of the U.S. electrical code since 1918). That's so electricity doesn't get into something you might touch, like the faucet in the bathtub, and cause you an electrical shock.

Your electrical power is supposed to return to its source strictly via the service cable or service drop attached to your home. The problem is, because your electrical system is grounded to your water main, those neutral or return electric currents can hitchhike along your water system (presuming the pipes are metal, not PVC, a nonconducting material which will stop them dead). That alternate path can take them all through your house and even to your neighbor's home. Likewise, the currents from your neighbor's home can enter yours. The electricity ultimately finds its way back to the power company via many possible paths. What makes these currents such unwelcome visitors is that when neutral currents take the road less traveled, they are not opposed by an equal but opposite current, creating those net currents we just talked about and likewise creating some large magnetic fields inside your home. And probably your neighbors' homes too.

So, take some measurements outside your house near the sidewalk or wherever your water main is located, because you may find a higher field there. Then take readings right next to where the water pipes enter your house. It's important to know if your water main is producing a high field, because there's a solution that *won't* involve the nightmare you're now imagining—replacing all your metal water pipes.

LET'S SEE WHAT WE FIND
INSIDE YOUR HOUSE

Inside, you should walk all around your home, making note of the places where you get high readings. First, check your electrical main panel—that might be in the basement or on the side of the garage if there is no basement—and the electric meter, usually on the outside wall. Look for the spot where your neutral wire is grounded to the plumbing, probably somewhere near a fuse or circuit breaker box. Your electrician calls it a grounding electrode conductor, which sounds like it's an intricate device, but it will probably look like a metal clamp holding a bare copper wire to a cold water pipe. You want readings there because, along with the readings you take outside around your water main, it will tell you if significant electricity is hitching a ride out of your house via the plumbing. After making your measurements, turn off the main circuit breaker and remeasure the fields. If they are unchanged, there is current flow into the house from outside on those parallel neutral paths and you will need to get professional help to interrupt the flow of current on these parallel neutral paths by installing nonconductive sections as discussed later. If they are gone, then there are wiring errors in the house and these will need to be corrected. This too will require professional help.

If the magnetic field level goes up or down, but does not disappear, there may be a combination of internal wiring errors and current flow from the outside via those parallel neutral paths. The next step is to break the current flow in the parallel neutral paths. After this, you tackle fixing the internal wiring errors.

Pay special attention to all the areas around your electrical outlets and appliances—on both sides of each wall. Walls don't stop magnetic fields. Sources of EMFs to be aware of as you position your furniture include transformers for electronic gadgets (the box at the end of the cord), televisions, computers, electric clocks, fluorescent and halogen lights, your computer, and your electrical service panel and outside electrical meter. The electrical cables in the wall can have net current due to wiring errors. Because net currents will produce magnetic fields over large areas, if you have them in your home, they may be from this type of source. In a 1995 study, leading EMF mitigator Karl Riley, taking readings from one hundred fifty buildings, found that 66 percent of the high fields were related to wiring and grounding problems. An amazing 65 percent of them were due to wiring violations.

Is this inspection really necessary? Not all of it. Unless you live in a tent in remote Alaska with no electricity, you can assume there are some EMFs in your home and do some simple measures to eliminate your exposure, which we dealt with in chapter 5, like arranging beds and other furniture where you spend a lot of time so they're as far away from electrical sources as possible (including fluorescent lights in the ceiling below bedrooms and computers, appliances, or your electric meter on the other side of walls). If you can find one in your area, you can also call an EMF mitigation firm that will come in, do all the readings for you, and recommend and implement fixes.

Be forewarned: There *are* some problems that aren't appropriate for amateur DIY solutions. And EMF mitigation companies aren't as common as plumbers and electricians, so you'll need to be savvy to hire the right kind of help you'll need.

BRING IN THE RIGHT ELECTRICIAN

If you can't find an EMF mitigator near you, you'll need to find an electrician you can work with. "You're probably not going to find someone who knows anything about electromagnetic fields," says Charles Keen, principle of EMF Services, a mitigation firm that has clients nationwide.

"You want to select an electrician who is a craftsman, who cares about the quality of his work, and is open to learning new things and solving problems," says Keen, a certified electrical inspector based in Florida, and a member of the National Electromagnetic Field Testing Association.

You may find that person in a larger franchise or nationwide firm, but you're more likely to be successful with a local, independent electrical contractor—someone who has been in the community for a long time

Finding Mr. Right

In 2001, at the age of forty-six, "Debbie Roberts" (she asked that I not use her real name) was diagnosed with chronic lymphocytic leukemia, an incurable disease that can be managed like diabetes for many years. After the initial shock and depression with receiving this diagnosis, she began looking for ways to keep herself healthy. Today, she eats only macrobiotic and organic food and feels "better than I did before I was diagnosed."

A few years after she learned she had leukemia, she discovered that for more than twenty years, the bedroom where she and her husband slept had EMF readings two to three times higher than what's considered safe. Her bathroom and the home office where she worked every day were also high. "I was literally bathing in a known carcinogen day and night for two decades," she says. "No wonder I was ill."

Some of it was emanating from a power pole about six feet from Debbie's bedroom window. Some came from her electrical and water system. But bringing those numbers down was not easy, even though she spent thousands of dollars to hire a well-known expert to test her home and write a report. The expert didn't do mitigation, so she had to find a local electrician to do it. Locating one who could do the job proved to be more difficult than finding a soul mate.

"The first one I called said, 'Hey lady, I've been doing this for thirty-nine years, and I'm fine.' He thought I was crazy. I said, 'Yeah, you're okay, good for you. Some people smoke for years and don't get cancer. But others do. It's not that simple.'"

and has a small shop. "You need someone who is scrupulous about code compliance," Keen says. Just bringing your electrical wiring up to code can eliminate many of your EMF problems. In fact, he says, what your electrician is going to spend the most time on is "fixing problems another electrician caused."

There may be dozens of such problems, but there are several that are extremely common, says Keen. They include:

Shared neutrals (or neutral-to-neutral wiring) and neutral-to-ground shorts. As I've noted, a safe electrical system is set up so that a wire that carries power to a lamp or appliance (a hot wire) sends the

Another electrician she encountered through a personal connection "treated me like I was a Ghostbuster—like it was some kind of joke." A third was so dismissive, Debbie's husband hung up on him.

Then they found him. "Ed was an eighty-something retired guy with an open mind. He went up and down our basement stairs a dozen times, and his son came to help him," says Debbie.

With the help of a plumber who asked to borrow our Gauss meter "to see if his home was safe," they installed a plastic piece that blocks the flow of electricity. Ed made a few wiring changes that helped bring the readings in the Roberts' house ultimately down to below 0.5 mG. (They still needed extra help: The cable company installed a device called an in-line isolator to prevent EMFs from the cable from beaming into her bedroom. When Debbie showed her report to the local utility company, they sent a crew out to rewire her entire neighborhood.)

In all, it took "nearly four months, two electricians, three cable company visits, two phone company visits, a plumber, and an EMF expert we found on the web, who flew in from Florida, to help us finally cure our house of this problem. We were so grateful to Ed," she says. "I still send him Christmas cards. He took on the job when the other electricians wouldn't."

The moral here: Don't expect to find Mr. Right on the very first call. Keep trying until you do.

current back via a second wire (usually white) called a neutral, running closely beside it in the same cord or cable, so any magnetic field it produces is small and localized. If that second wire is farther away or carrying a different load, it can create an imbalance, which translates into a magnetic field with greater reach. A Gauss meter reading will tell you if any of the wires in your walls have this net current—you'll see a jump in the magnetic field.

You have many circuit breakers in your electrical panel. These power what are termed branch circuits. If two or more of these branch circuits happen to meet up in the same electrical junction box and all the white or neutral wires are connected together, you will have unbalanced current flow in the cables. So whenever there is current flow on any of these branch circuits, you will have an elevated magnetic field in all the spaces where these cable are running, which can make it tough for your electrician to actually identify the source of the fields.

If a neutral wire is connected to a ground wire anywhere but in the main electrical panel, you have another source of elevated magnetic fields because you don't have current flowing in the opposite direction. Your neutral wires should never be attached to anything grounded—a ground wire, a grounded electrical box, or the frame of a grounded appliance. Otherwise, the return or neutral current will take that alternate path, destroying the natural magnetic field cancellation in the cable, generating a far-reaching magnetic field.

Another potential source of elevated magnetic fields occurs when an electrician incorrectly wires a three-way switch—one that controls light from more than one location—by sourcing the power supply (hot) and power return (neutral) wires from different branch circuits in different switch boxes. This causes unopposed current flow on a single wire and as we have already talked about, high magnetic fields in the rooms near the wires supplying these two circuits as well as the wire that runs between the two switches. This is a violation of the National Electrical Code, but more important for your well-being, having the wires on different walls creates a huge magnetic field between the two wires, which could potentially be an entire room. Your electrician can remedy all these problems by rewiring. (Be wary of anyone who says your entire house needs to be rewired. Most of these EMF problems are caused by one or more wiring errors that tend to be localized. A shoddy electrician or a weekend handyman is likely at fault.)

If you live in an urban high-rise or suburban apartment and get high magnetic field readings, look for the building's transformers and switching boxes, which can create high magnetic fields. Additionally, it is common for current to be running on the steel beams in high-rise buildings due to connections between neutral and ground, placing neutral current flow on the grounding system. If your building owner is reluctant to pay for mitigation, you will have to move furniture such as couches, chairs, and beds to an area of your apartment where the readings are lower.

Water line problems. A good electrician will probably also tell you that a plumber needs to be involved because your electrical system is grounded to your water main for safety which, as we've seen, can generate a large magnetic field. The usual solution is something called a nonconductive (meaning it doesn't conduct electricity) section of plastic pipe that will eliminate plumbing currents by breaking the metallic path on which current flows. But it's not a simple job. It's usually fitted to underground pipes, so backhoes may be part of this solution. Make sure that these changes are satisfactory to your local plumbing and electrical inspection authorities. Another remedy suggested by some mitigation experts is to have your plumber replace copper connectors to all the appliances hooked up to your water system with plastic or nylon piping to avoid routing return power to your washer, ice-making fridge, or even your sink. (Current can also hitchhike along gas pipes inside the house, though inside they have a plastic coupler located near the meter to prevent current flow. It may also travel along TV or phone cables. If you pick up a strong field near your gas pipes, contact your local utility, which will likely fix that for free. Talk to your cable TV and telephone service provider as well.)

Knob and tube wiring. If you live in a home built in the 1940s or before, you may still have the old-fashioned wiring system in which single, insulated copper conductor wires are run within the hollow areas of a wall or ceiling cavities (they need some air space, so heat has a chance to dissipate). They pass through drill holes in joists and studs in porcelain insulating tubes, supported by cylindrical porcelain knob insulators to keep them from touching the wood. Check your basement and attic—the knobs and tubes are white. Because the flow and return current wires can be separated by six to sixteen inches, a large magnetic field and electric field is created.

Even if you have new wiring in an old building, that doesn't mean there aren't some remnants of the old knob and tube wiring still inside the walls. It can cost thousands to replace the antiquated system, but it's likely to be worth it for more reasons than to reduce your EMF exposure. Designed to carry only a modest electric load, knob and tube wiring is usually not up to handling our twenty-first-century plethora of appliances and electronics, so it can be a serious fire hazard. Two more reasons to replace it: with knob and tube wiring, you can't add insulation to walls or the attic because of fire danger, and some insurance companies won't issue homeowner's insurance when it's present, even if only in part of the house.

Dimmer switches. They seem so innocuous, but believe it or not, says Keen, dimmers use an electronic transformer that produces electric fields in the RF end of the spectrum. (Your AM radio will pick it up as static.) The problem is the wire that connects the dimmer to the electrical panel and the dimmer to the light acts like an antenna broadcasting this RF into the house. "The emissions are actually greatest in the middle dimmer position," Keen says. "You're bathing yourself in a strange combination of emissions that span a broad frequency range." You can have the dimmer switch removed (recommended), or ask your electrician about a de-buzzing coil, which was developed to remove the buzzing that you frequently hear in the switch (or even in some lamps as the filaments vibrate) but which also reduces the EMF.

Power lines and substations. Not even the best electrician can help you with this one. You have little or no control over the overhead or underground power lines or over substations. Power lines don't carry a uniform load of electricity; it varies daily, even hourly. That often rapidly changing load can create magnetic fields, "and a power line has nothing to cancel it," says Keen.

Distance is your best remedy. While under the line, electrical fields can range from a few volts to 200 V per meter and magnetic fields from 10 mG to as much as 70 mG during peak times. While fields diminish in strength the farther away you are from the lines, fields can still be measured at one-quarter to one-half mile from transmission lines, according to Gust. Likewise, magnetic fields around the lines outside power substations and even power transmission lines—the big ones—can be detected but diminish with distance.

Interestingly, magnetic fields close to some typical electrical appliances that you have in your home—like hair dryers, electric shavers, vacuums, and electric ranges—have stronger magnetic fields than power lines! The good news—you have more control over your appliances and how long you are exposed to them. Even if you can't afford to move away from power lines or substations, you *can* minimize your exposure elsewhere—and you should, particularly if you have symptoms of electro-sensitivity (see chapter 3).

If you have the money—and the remedy can cost anywhere from five thousand to twenty thousand dollars—you can install what's called active shielding or active magnetic cancellation, a system that monitors the power line field and adjusts to compensate for the variation of loads the line carries—essentially creating that equal and opposing field of energy that neutralizes the magnetic field the power line generates. Keen's company installs them commercially, and if you have enough space and money and a great need, the unattractive wires and coils can be placed underground. It's the same kind of system used in scientific laboratories, where a magnetic field can interfere with sensitive equipment.

Some of the simplest of these solutions can help you when your electrical power becomes polluted by high frequency distortions called transients and microsurges, a common condition called dirty electricity. Also, be aware of cell towers and antennas in your neighborhood. They have been multiplying like crazy over the past fifteen years! To gauge the risk of your particular locale you can go to the BRAG Antenna Ranking of Schools Report and find the section that teaches you how to grade any location, school or otherwise. This report is located on www.magdahavas.com.

In the next chapter, find out about this phenomenon that's the result of, among other things, our antiquated electrical grid and the proliferation of electronics in our homes.

THE NEWEST ZAPPER: DIRTY ELECTRICITY

In the 1971 film classic, Clint Eastwood's character is called "Dirty" Harry because he always does a good job, but he also has a bad streak. Or as the tagline of the movie reports: "You don't assign him to murder cases. You just turn him loose."

And if you think about it, electricity works much the same way. We turn it loose everywhere we go; from home to car to office to mall, and somehow, almost miraculously, it manages to make all the lights go on, make all the gadgets and gizmos run, and do just about anything we ask of it. But as you will soon find out, while electricity is doing all of our bidding for us, it also stirs up quite an invisible and destructive mess at the same time.

Dirty electricity has become such a pervasive problem that I was compelled to devote an entire chapter to this emerging electropollution health threat. Dirty Harry . . . meet Dirty Electricity.

HOW DOES ELECTRICITY
GET DIRTY?

Did you know that the electricity that powers up coffeemakers, lights streets, and keeps commuter trains running all over the country travels on almost one hundred sixty thousand miles of high voltage transmission lines? Called the grid, it is so antiquated, it costs U.S. industry billions of dollars because of micro power surges that act like rust on expensive machinery, gradually wearing down the machinery until it fails.

And it can be wearing on the human body, as well, which you will see in a moment.

The U.S. electrical grid is polluted by what's called invisible dirt—high and mid-frequency signals—that may be responsible not only for damaged machinery, but also a host of physical symptoms in people, from headaches and joint pain to memory loss and seizures.

Here's how dirty electricity happens: Power is transmitted to your city via high voltage transmission lines that run on those erector-set looking towers. The voltage is very high, 66,000 to 765,000 V. These high voltage transmission lines take power from its source—a coal, nuclear, or natural gas plant—to the vicinity of your city, where it terminates at a substation. In the substation, the high voltage is lowered to 4,000 to 69,000 V in preparation to be sent to homes and businesses via the local distribution lines. This voltage is much too high to be safe in a building, so when the power reaches your neighborhood, it's stepped down to 120 to 240 V by those barrel-shaped transformers you see on power poles, and then it's fed into your home or business. If you had a device called an oscilloscope and you had looked at the voltage in your home's power system back in 1950, you would have seen a smooth wave form that looked sinusoidal— like a gently undulating wave in the ocean.

WHERE THE DIRT COMES FROM

Many years ago, before the discovery of solid state devices now found in all electronic equipment, the electricity that powers our homes, buildings, and factories was like a meadow in the country, gently varying, quiet, harmonious, and clean. Today our electricity is no longer any of those things. It is contaminated by high-frequency transients and harmonics described as dirty electricity or dirty power. This dirt is much like noisy static you might hear on a radio playing wonderful classical music. The underlying music is there, and it could be beautiful and relaxing, but all that static is too irritating, so you change the station or turn off the radio.

Unfortunately, you can't turn off dirty electricity. It is everywhere in our environment. You are generating this dirt in your very own house, as are your neighbors and the office down the block and the factory in the industrial part of town.

The dirt is produced by the workings of all our electronic equipment such as computers, TVs, radios, microwave ovens, compact fluorescent bulbs, light dimmers, digital clocks, and cell phone chargers.

The dirt follows the electrical lines around and out of your house following the wires that supply electricity to your home. The neighbor's dirt, the office's dirt, and the factory's dirt follow the same paths and end up in every house, office, and factory.

We have assumed that this form of energy is not biologically active, meaning it does not affect living systems. However, research tells us that this dirty electricity may indeed be biologically active and potentially responsible for an array of symptoms.[1] "When filters to reduce the dirty electricity are installed in homes and schools, symptoms such as chronic fatigue, depression, headaches, body aches and pains, ringing in the ears, dizziness, diabetes, impaired sleep, memory loss, and confusion are reduced," says Gust, who does indoor environmental mitigation. Says Gust: "Based on this early research, it is estimated that a high percentage of the population may be sensitive to dirty electricity [although the figure generally bandied about in scientific circles is 2 percent]; children may be more sensitive than adults, and dirty electricity in schools may be interfering with education and possibly contributing to disruptive behavior associated with attention deficit disorder; dirty electricity may elevate plasma glucose levels among diabetics, and exacerbate symptoms for those with multiple sclerosis and tinnitus."

FELLED BY CURRENT

In 1996, as a twenty-three-year-old newlywed, Catherine Kleiber moved with her new husband, Dan, to a farm near Waterloo, Wisconsin. Six months later, she was wracked by a series of mysterious symptoms. "I started having problems with muscle weakness, shortness of breath, dizziness, lightheadedness, chills and hot flashes, and problems with the circulation to my extremities," says Catherine who, as a student at the University of Wisconsin-Madison, loved "booking from one end of the campus to the other. I was a very active person, and the daughter of two doctors who never saw doctors because I didn't get sick."

But by the winter of her first year of marriage, she was severely ill. "I could hardly walk up the stairs. When I got to the top, I would have to bend over because I was so lightheaded, I thought I would pass out. My heart would go, *boom, boom, boom*. I had acid reflux so bad I couldn't sleep at night. When I woke up, I would feel like I was hit by a Mack truck and then run over by a train."

She went from doctor to doctor and was eventually diagnosed with chronic fatigue syndrome. Since she "didn't much like that diagnosis because I didn't think my body just decided one day to self-destruct for no reason," she began doing research on her symptoms to see if there might be another explanation. That's when she came across an article on the problem of dirty electricity in rural areas. "The story mentioned that one of the hallmarks of a farm that had been polluted by dirty electricity is that the farmer had stopped dairy farming because milk production dropped off or the farmer himself got sick. Our farm hadn't been a dairy farm since the 1970s."

So she called the editor of the agricultural magazine in which she read the piece. "Just about every doctor I saw thought I was nuttier than a fruitcake and that my condition was psychosomatic. Instead of that response, the editor said, 'Oh my goodness. You have to call the electrical quality expert Dave Stetzer and get him out to your place. He knows about this, and he has a solution.' I was nearly in tears because someone finally took me seriously."

Stetzer, CEO of Stetzer Electric Company in Blair, Wisconsin, visited the Kleibers' farm and took measurements. The farm, it turned out, was so close to a transformer that the Kleiber home was the path of least resistance for the return current, which tends to pick up dirty electricity as it moves around the grid. Stetzer showed Catherine a reading he took of her as she stood at her kitchen sink. "I realized then why capping strawberries at the sink made me dizzy and gave me heart palpitations," she says. "There was this odd-looking wave form going right through my body." It appeared that not only was her water system carrying dirty electricity (some of her neighbors got electrical shocks just touching their sinks), her entire house was buzzing with it. It was the EMF version of *The Amityville Horror*.

"We're at the end of the [power] line, and that's where high frequencies like to get off," explains Catherine, who now manages the educational website www.electricalpollution.com. And, like most Midwest farms, theirs boasted rich dark fertile soil—highly conductive to the excess current that spilled into the earth. "One mile away is the substation that was at one time the substation for Waterloo [North Hydro], so we're right in the path of access of 70 percent of the current returning through the earth. If you could see it, it would probably look like a stampede, but you can't see it. But if I would sit on the ground I would start hurting."

Before Catherine moved to her farm, she had never experienced electrosensitivity. Today, even though she has taken extreme measures to protect herself—installed filters that block the high frequency dirt, discarded most of her electrical equipment (including TV and computers), shuts off the electric power at night before she goes to bed—her overexposure to ground current seems to have permanently sensitized her. Even a small exposure to a cordless phone or a friend's Wi-Fi can affect her.

"Once you're overexposed, you don't get to turn back," she told me. "If I'm in a good environment, I'm fine. I have some lingering damage in that I still have trouble with the circulation to my extremities. My autonomic nervous system must have suffered some damage. I don't feel quite as sharp as I used to be."

And, as the mother of two young sons, she admits it upsets her that her family's life has become so circumscribed by her condition. "We can't go to people's houses. We haven't gone out to eat in years. I love Dan's family, but when you get seventy people together and they all have cell phones, it can be unbearable for me. When I take the boys to gymnastics class, I have to call ahead to make sure they turn their Wi-Fi off. I have a lifelong friend whom I haven't seen even though her husband is dying, just because their home is wireless.

"This is what I tell people," Catherine says. "You don't want to be where I am. The best way for you to not be where I am is not to overexpose yourself. And if you are where I am, the only way to get better is to create a safe environment."

UNZAP YOUR HOME WITH THE 98 PERCENT SOLUTION

Dave Stetzer, the electrical quality expert who took measurements at the Kleibers' Wisconsin farm, has made a career out of creating safe environments, often at his own peril.

"When I'm out in the field, I'll get heart palpitations and chest pains. It hurts down my arm—all the symptoms of a heart attack," he admits. His blood sugar also soars.[2]

But when he gets back to the office, which he fitted with high frequency filters of his own design, he starts to feel normal again.

Stetzer, along with Martin Graham, Ph.D., professor emeritus in the College of Engineering at the University of California at Berkeley, developed a filter that can reduce the amplitude of these higher frequency voltage surges (they're also called microsurges) within the range of 4 to 100 kHz. They're not unlike the power surge strip you have for your computer. Stetzer and Graham also created a microsurge meter to measure the extent of the problem to help determine how many filters an individual home or business needs. (Graham actually holds the patent for the device that measures electrical pollution on a power line.) The filters and microsurge detector are small enough to plug into a normal electrical outlet. Stetzer estimates that the average home needs about twenty filters, called Graham-Stetzer filters or Stetzerizers, to neutralize dirty electricity (cost is $35, so it is $700 for the entire house, see resources).

He and Graham created their own solution when it became clear that the electric power industry (a collection of about five hundred utility companies) wasn't going to implement a system-wide remedy—a project that could potentially cost billions per utility. (The bill for one relatively small upgrade in Kansas alone was estimated at $2.2 million per mile, a cost that would likely be passed on to customers.)

"By 1980, dirty electricity was such a problem that buildings were burning and equipment was failing," Stetzer says. "The IEEE [an acronym for the Institute of Electrical and Electronics Engineers] adopted standards on how to deal with it. The utility is supposed to put up bigger wires to be able to handle this. Then they have to put filters on your system because the high frequencies can get into the RF range. Once they're on the wires, the high frequencies can radiate like a horizontal antenna. If you work in a metal building, for instance, it will suck it in like a satellite dish."[3]

But those standards have never been met—hence, Stetzer's seven-hundred-dollar fix, which he says works "about 98 percent of the time."

A SECRET STUDY—HELPING CHILDREN WITH LEARNING DISABILITIES

Stetzerizer results are usually fast and so amazing that researchers like Magda Havas, B.Sc., Ph.D., associate professor of environmental and resource studies at Trent University in Ontario, are still "blown away" each time they conduct a study.

Havas admits that she was skeptical about the filters when she got a call from a mother whose hypersensitive diabetic daughter could no longer attend a Toronto private school for children with learning disabilities: At the end of the school day, the teenager was so fatigued, she would have to go to bed the minute she came home. The woman had convinced the principal to allow her to install Stetzerizer filters at the school at her own expense and wanted Havas to do a study there. The principal had agreed to the study.

"I do research on electromagnetic pollution, but I didn't know if it affected people. This was a new take on it for me," Havas admits. "The woman had installed [the filters] at home and was convinced they worked. Everyone in the house seemed to benefit. Even the dog was better behaved." Havas laughs. "I was very skeptical. I didn't think they could do anything for human health at all."

No one except the principal knew the purpose of the study. Employees were asked to fill out daily questionnaires over six weeks—three weeks without filters, three weeks with. Since the school population was small (only fifty people) and only about 2 percent could be expected to be hypersensitive, Havas thought she might see an effect in two people at the most. But when the results came in, she was stunned. "I thought I'd made a mistake. I recalculated many times. Instead of 2 percent, 55 percent improved. I thought, 'This is unheard of. It can't happen.'" Yet even those who thought they didn't feel bad before felt better. "We did this between January and March—and February in this part of the world is depression haven. In February, we haven't had any sun for a long time. But during the month of February when the filters were installed, these people were feeling incredibly good compared to January and March [when they removed the filters]."

Teachers reported feeling less tired, frustrated, and irritable; students, especially those in elementary grades, were less likely to be late to class, less disruptive, and more able to focus. In fact, some symptoms of ADD and ADHD improved after the filters were installed.[4] A later, similar study at a Minnesota school achieved comparable results.[5]

In subsequent small studies of the effects of the Stetzerizer filters on other populations, Havas has found that they also improve blood sugar and reduce insulin needs in both type 1 and type 2 diabetics and relieve some symptoms in people with multiple sclerosis.[6] In the case of diabetics, she speculates that the link may be stress, which raises blood sugar. Studies have found that the body produces stress proteins in reaction to EMF exposure, so if the exposure is lessened, Havas theorizes, then blood sugar will drop.

People with multiple sclerosis (MS) have a variety of symptoms, from blurred vision to paralysis, that are caused by damage done to the myelin sheath of their nerves. "The myelin sheath is there to insulate the nerve against extraneous impulses, like a cord to a lamp," explains Havas. "When you start putting holes in it, the signal doesn't travel as quickly. The brain says, 'Take a step,' and there's a delayed reaction. The nerves are very sensitive to electromagnetic influence. They use electromagnetic energy to communicate, what Dr. Ross Adey called cells 'whispering' to each other. Electromagnetic fields from outside the body can interfere with that communication, which is already impaired in people with MS."

In one case Havas documented, a woman with MS was able to walk without a cane for the first time in years once she installed the filters in her home.

AVOID BEING ZAPPED AT HOME

While most people seem to benefit from Stetzerizing their homes, for others, like Catherine Kleiber, it might not be enough. In addition to following the tips in chapters 3, 5, and 6, take these steps to reduce RF exposure in your home. Bring your electrician or an environmental expert in to inspect your entire house, inside and out, and correct any problems. You might want to have the book *Tracing EMFs in Building Wiring and Grounding* by Karl Riley for your service person to consult.[7] Riley literally wrote the book for electricians on mitigating wiring errors that promote EMFs. Some of the problems your electrician should be looking for will include:

- Electrical code violations. One common violation, according to Riley, is joining the two neutral wires—the ones that return electricity to its source—from two branch circuits, one of which may light one part of the house, while the other lights another. If you flick on a switch in one room, the return current will flow back to your panel in the basement through *both* neutral wires. That creates an imbalance in the amount of current flowing to and from the source inside the supply cables (a net current). Remember, when currents are identical, the magnetic field is nearly canceled out. When they're not, a magnetic field exists between any cables that have a net current.

- Loose or poor connections in your electrical system that may be arcing.

- Connections between the neutral wire and the electrical ground that divert the neutral return current to another path, such as water pipes, gas pipes, heating ducts, HVAC freon lines, the grids for dropped ceiling panels, metal studs, or conceivably any metal in your home.

- Switches that are old, worn, or poorly made that may be arcing.

- Neutral wires whose insulation has been pinched or punctured, causing an inadvertent connection between neutral and ground wire.

- Tree branches that scrape or bump overhead wires (your utility may trim them for you).

- Dimmer switches.

If you have an EMF mitigator measure the fields in your house and he finds that AC magnetic fields are coming from a problem caused by the power company's local distribution system, call or send the report to your local utility engineering department. I spoke to one woman whose home had high EMF levels when the power was shut down, indicating a problem coming from outside the home. When she sent the report to her electric company, she said, they "came out and rewired the entire neighborhood." Now, her home measures around 0.5 mG all the time.

Talk to your electrician about harmonic filters or line/power conditioners, which are recommended for the protection of expensive electronics such as home theater systems. They basically suppress voltage spikes coming from the utility power system, and as an unintended plus they clean up dirty electricity in some frequency ranges—filtering out transients and harmonic distortions caused by RF interference. For about ten dollars, you can also install an in-line ground isolator that cleans up any dirty electricity on your outdoor TV cable.

Contact your utility company to check the power lines that run into your home to make sure all connections are secure. Some may be so old, worn, and corroded that the flow of return current is impeded.

Install an RF filter on your phone line, which can also bring unwanted RF into your home. You can buy them for under twenty dollars at Radio-Shack and other electronics stores. They're designed to reduce high frequency white noise on phone lines, which pick them up via the air or on electric utility ground currents. Shielded phone wire available from EMF catalogs and websites such as www.lessemf.com may also help.

In the next chapter, I'm going to tell you about what may be the single most dangerous device that virtually all of us use on a frequent basis. This may be the most inconvenient truth thus far.

ZAP-PROOF
YOUR PHONE

So, what are you going to do about your wireless phone? It's time to focus on the one thing that everyone seems so hesitant to address—and for pretty good reasons, as you will soon find out. But please don't skip over this chapter. Just rest assured that the one thing I won't be telling you to do is gather all your family's cell phones and bury them in a very deep hole in your backyard.

Absurd, right?

But here's what can and will happen after you read this chapter. You will gain enough understanding of why cell phones can be hazardous to your health and how to begin to use the simple precautions I recommend at the end of the chapter to minimize exposure to the electromagnetic radiation while still using your cell phones as necessary.

Fair enough? Let's get started.

A few years ago, I read a newspaper column about one of the un-imagined consequences of the new walk-and-talk culture spawned by cell phones. The writer, who was blind, pointed out that the sight-impaired, who rely on their sense of hearing, are assaulted constantly by the never-ending conversation on urban streets, often feeling foolish be-cause they've responded to the "Hello, how are you?" that *wasn't* meant for them.

It made me think about the ever-expanding presence of cell phones and my first startled reactions when I would see someone walking down the street, hands gesturing, talking out loud to what I thought at first was an imaginary companion—before, of course, I saw the telltale earbud.

Imagine for a moment that those conversations we hear on an average city street are microwaves. If you're within four to five feet of someone talking on a cell phone or just carrying a switched-on cell phone—without even talking to someone yourself—you *are* being bombarded by plumelike waves of microwaves, similar to secondhand smoke. Those energy waves emanate from all sides of the cell phone's antenna that picks up signals from and sends signals to a base station or antenna—one of those nearly quarter million microwave-emitting cell phone antennas that dot the U.S. landscape, some camouflaged as trees, church steeples, or grain silos, others perched on the roofs of schools and hospitals.

The closer you are to any cell phone—in an elevator, on a crowded bus, or with your own cell phone held up to your ear or tucked into your pants pocket—the stronger the signal that reaches your brain or other organs. In fact, the wireless industry's own research has found that the microwaves from a cell phone penetrate about two inches into an adult brain, deeper in a child's head because the immature skull is smaller and thinner. As an acknowledgment that the RF radiation field is absorbed by the human body, cell phone manufacturers list the SAR—specific absorption rate—for cell phones in each phone's instruction manual. SAR is the measurement of the strength of the radiation field your body absorbs. Cell phones have been around since the 1980s, though the earlier models weren't the sleek, palm-size phones and PDAs on the market now, with their Swiss Army knife capabilities: not only can you call your friends in Europe from your home in Ohio, you can surf the Internet, read your e-mail, get directions, watch a movie, listen to music, play a video game, calculate a tip, count your calories, take a photo, make a film, even bid on eBay. This de rigueur accessory is based on communications technology developed for the U.S. Department of Defense. And although cell phones radiate microwaves—similar in some ways to your microwave oven, though they still can't defrost a frozen dinner—the telecom industry convinced the FDA to allow them to be sold without any premarket testing on the grounds that the power they emit is far less than the high power used by microwave ovens, which is strong enough to heat human tissue.

In fact, the first studies of cell phone safety were prompted by Congress in 1993, alarmed by the publicity surrounding a lawsuit against one phone manufacturer by a Florida man whose wife, Susan Reynard, died of a brain tumor which was on the same side of her head where she held the phone, chatting for hours a day.[1]

So the Telecommunications Industry Association set up a nonprofit organization, Wireless Technology Research (WTR), and commissioned a twenty-eight-million-dollar study into cell phone safety, tapping to lead it an epidemiologist named George Carlo, Ph.D., J.D., who was fresh off a project with the FDA, investigating the safety of silicone breast implants. In his book *Cell Phones: Invisible Hazards in the Wireless Age,* Carlo writes that his first inkling that something might be amiss with his new project was the claim the industry was making in its PR push that there were "thousands of studies" proving cell phone safety—and Carlo couldn't find one.[2]

Sensitive to the possibility that he might be seen as a shill for the industry, he recruited a host of prominent scientists to work with him and established a peer review committee to oversee their work. Since there were no studies on cell phone safety, Carlo and his colleagues had to develop novel protocols for testing exposures on cells and laboratory animals that were equivalent to head exposure in people using cell phones. Over the next few years, what they found in their many studies was DNA damage (broken strands of DNA called micronuclei are found in the blood and can lead to tumor formation), impaired ability of cells to repair themselves, leakage of the blood-brain barrier, and interference with cardiac pacemakers, results later confirmed by other studies both in the United States and abroad. Needless to say, the telecom industry quickly distanced itself from its own research and from Carlo, slashing funding and stopping public dissemination of the findings, although Carlo and his fellow scientists did bring the results of the studies to light on television, and Carlo eventually in his book.[3]

Carlo's nearly suppressed work became the match that lit the fire. To date, there have been a multitude of studies on cell phone safety, including Interphone, a thirteen-country study. They're far from conclusive, but

Is There a Tower or Antenna Near You?

The website www.antennasearch.com will give you a list of all the registered cell towers, antennas, and tower applications near your home. The free site contains information on more than 1.9 million such structures in the United States. There is evidence of an increase in symptoms from fatigue to leukemia in people living within a quarter mile of cell and broadcast towers.

a look at funding sources may help you understand one reason why: in one study correlating results to funding, industry-funded research was far more likely to show no effect (72 percent versus 28 percent) than independently funded studies (67 percent suggested an effect, 33 percent no effect).[4]

While some of this research has linked cell phone use to everything from low sperm count to epilepsy to Alzheimer's disease, the main focus has been on brain tumors. In 2008, Ronald B. Herberman, M.D., director of the University of Pittsburgh Cancer Institute, garnered headlines and appeared on *Larry King Live* after issuing a memo to his staff urging them to take stringent precautions when using their cell phones because of the risk of developing brain cancer.

"Although the evidence is still controversial," wrote Herberman, "I am convinced that there are sufficient data to warrant issuing an advisory to share some precautionary advice on cell phone use." He counseled his staff to limit the amount of time they spent on their cell phones and to avoid making calls in public places "where you can passively expose others to your phone's electromagnetic fields."[5]

What so alarmed Herberman? After all, most of the major cancer and health organizations in the United States, while expressing some caution, seem to believe that the evidence suggesting a problem is slim. What Herberman knew is that it's not nonexistent, and the few studies that have looked at heavy (1,640 cumulative hours of calls) and long-term (ten years or more) use have been fairly consistent in their findings. Consider:

- In Sweden, where mobile phones have been in use since the early 1990s (it's the home of Nokia and Ericsson), the Swedish National Institute for Working Life examined the cell phone habits of more than nine hundred people who had been diagnosed with brain tumors and found that those who used the phones most often and for the longest period of time had a 240 percent increased risk for a malignant tumor on the side of the head where they typically held the phone.[6] (Like the Florida woman whose brain cancer death led to the first lawsuit in the United States.) The results conflicted with a British study released about a year before it, but the UK researchers cautioned against drawing any definitive conclusions about cell phone safety from their work. "The results of our study suggest there is no substantial risk in the first decade after starting use," said lead researcher Anthony Swerdlow of the Institute of Cancer Research in published reports. "Whether there are longer-term risks remains unknown, reflecting the fact that this is a relatively recent technology."[7]

- In 2009, a multinational study found that people who use mobile phones before the age of twenty are more than five times as likely to develop a malignant brain tumor. Swedish oncologist Lennart Hardell, M.D., lead author, also noted that people who start using a cell phone later in life have a 1.5 greater risk of malignant brain tumor than the general population. Hardell has also found increased risk in people who use cordless phones for more than ten years.[8]

- A study by a Mayo Clinic–trained neurologist also found a strong link between duration of cell phone use and brain tumors. Vini G. Khurana, Ph.D., of the Canberra Hospital in Australia did not test the hypothesis itself. Instead, he did a meta-analysis—a review and synthesis of a number of studies—of all the available research that looked at cell phone use for ten years and more. What he found was that long-term use doubles the risk of brain tumors on the side of the head where the phone is usually held.[9]

THE SCIENCE BEHIND THE CANCER CONNECTION

How can RF exposure cause cancer? Like ELFs, RFs—even those below safety standards—have been shown to damage cellular DNA, which can lead to mutations that can cause cancer and can even be passed to successive generations. Like other EMFs, RFs are linked to the formation of free radicals which, as we learned in chapter 2, also damages cellular genetic material at a time when the body's own protective antioxidant production (melatonin, superoxide dismutase, glutathione) is being suppressed. In animal studies done by Henry Lai and his colleagues at the University of Washington, melatonin and a vitamin E analog (another antioxidant) blocked the effect of 60 Hz ELFs on the DNA of rat brain cells.[10] Lai also found that reducing iron (using a chelating chemical) also blocked the effects of the ELF, suggesting that iron—a magnetic substance—may be involved in the cellular genetic damage.[11]

Studies have also shown that low-level RF exposures trigger the body to produce heat stress proteins, part of the body's stress response to environmental toxins. While the process is meant to be protective, it can be harmful if the stress response becomes chronic.[12] There's also strong evidence that RF can cause a break in the blood-brain barrier that prevents toxins in the blood from reaching the brain.[13] This could allow environmental chemicals in the bloodstream—and the Centers for Disease Control and Prevention have found 148 of them, from lead and mercury to cancer-causing dioxins and materials like polychlorinated biphenyls (PCBs) used in transformers, in the average American—to enter the brain where they could cause nerve cell death and can trigger tumor formation.[14]

There have also been cancer clusters, including childhood leukemia and Hodgkin's lymphoma, among people who live near cell phone towers as well as military radar and communication bases. In fact, one Israeli study found that people living near a cell phone tower had a four-fold risk of many different kinds of cancer, including breast cancer, Hodgkin's disease, and cancers of the bone, kidney, ovary, and lung than that of the population as a whole. Women were disproportionately affected (seven out of eight cancer cases were women) and in the year after the study ended, eight more cancer cases were diagnosed.[15] I believe that women are more affected because their sympathetic nervous systems are simply more biologically attuned to the environment and can sense energy, perhaps due to the innate "mothering" instinct.

By far, the most common complaints from people who live near cell phone towers are sleep disruption, headaches, dizziness, depression, lack of concentration, and muscle fatigue—all symptoms experienced by radar and radio communications operators and people living in homes contaminated by dirty electricity, a condition called radio wave sickness (see chapter 3). The effects have been seen in people living less than one-third of a mile from the towers and even in those who didn't connect the proximity of the tower to their symptoms, which effectively eliminates the placebo effect.[16]

One leading theory on why RF causes seemingly unrelated conditions is the stress connection. These conditions are all symptomatic of stress-related disorders caused by an overstimulation of the sympathetic nervous system that elevates cortisol, which eventually leads to adrenal burnout and a dysfunction of the immune system. When the immune system is stressed, then a whole host of health disorders can emerge, such as short-term memory problems, irritability, anxiety, headaches, inability to concentrate, and heart irregularities.

In a fascinating exposé of the cell phone industry, *GQ Magazine* published an article in February 2010 by Christopher Ketcham, who wrote:

> The Telecommunications Act of 1996—a watershed for the cell phone industry—was the result, in part, of nearly fifty million dollars in political contributions and lobbying largesse from the telecom industry. The prize in the TCA for telecom companies branching into wireless was a rider known as Section 704, which specifically prohibits citizens and local governments from stopping placement of a cell tower due to health concerns. Section 704 was clear: There could be no litigation to oppose cell towers because the signals make you sick.[17]

Currently the government's intent to expand broadband coverage will mean "an even denser layer of radio frequency pollution on top of what has developed over the last two decades," says award-winning journalist B. Blake Levitt, whose groundbreaking book *Electromagnetic Fields: A Consumer's Guide to the Issues and How to Protect Ourselves* was re-released in 2007 after going out of print. Providing super Wi-Fi coverage for up to seventy-five square miles, "WIMAX will require many new antennas."

Consumers around the world have aggressively protested the march of Wi-Fi and cell towers. In countries like Ireland and Spain, they have taken action by bulldozing cell towers, while local citizens cheered them on. In 2007, a retired telecom worker in Australia hijacked a tank and rammed six cell towers to the ground before local police could stop him.

In America, in government circles, there is mostly deafening silence.

SO, WHAT CAN YOU DO TO PROTECT YOURSELF?

With cell phones themselves, consumers have considerably more control. That said, a 2004 Massachusetts Institute of Technology survey found that people have a love-hate relationship with their cell phones—nearly one in three say they're the one invention they can't live without. Like most modern-day conveniences, they have their benefits and social costs. Cell phones are literal lifelines in emergencies. One study found that emergency response times by ambulance crews improved because of cell phone use. There was also a scientific study that found that people who use cell phones while driving were four times more likely to be in accidents than people who weren't using cell phones. Cells have been a boon to business, allowing workers to be in contact with clients or the office from any locale, though they've resulted in a general blurring of the line between work and private/family time. In 2009, there were over 285 million cell phones in use in the United States; worldwide, more than 4 billion. I don't believe that cell phones—nor their towers—are ever really going to go away, even if they do pose a significant public health challenge from RF pollution. But, there's good news on the horizon.

How Do Cell Phones Work?

Cell phones operate in the region of the electromagnetic spectrum assigned to radio waves, between FM radio and infrared light. The frequency of the cell phone carrier wave is lower than that of a microwave oven or a wireless Internet connection and much, much lower than radar or satellite stations.

"The unique thing about information transfer is the carrier wave used to convey packets of digitized information," says Larry Gust. "Various aspects of the technology produce energy vibrations that are within the human cell's vibratory sensing capability. This portion of the signal is known as the Information Carrying Radio Wave (ICRW). You can think of the carrier wave as a clothesline on a pulley and the clothes pinned to the line are the information-containing packets."

Cell phones use a base station that receives and transmits radio waves. When you punch in the numbers for your spouse or the local pizza place, a signal is sent from your cell phone antenna to the base station antenna nearest you in the *cell,* which is what the effective radius of each strategically located base station is called. That base station assigns your call to one of the available channels. Then radio waves are simultaneously sent and received and the call goes to a switching center so it can be transferred to another carrier or cell phone. That allows you to warn your partner that you're coming home late from the office and to order a pizza for pickup. Your body's exposure from one cell phone call varies depending on your weight, how long you stay on the call, and how far away from a base station you are. In general, the farther away you are, the more power your phone needs to keep the connection, so your brain is absorbing more RF. The good news is that you can always turn your cell phone off after a short call.

You can significantly minimize your exposure and the secondhand RF "smoke" to which you expose others by doing just a few simple, common sense things:

Buy low. Choose a cell phone with a low SAR rating. SAR stands for specific absorption rate, which measures the strength of a magnetic field absorbed by the body. If you don't have your instruction manual, which is where that number should be (but isn't always), you can find SAR listings at the Federal Communications Commission website (www.fcc.gov/cgb/sar/) or you can request SAR information from the manufacturer or your carrier. You'll need your model and FCC ID number, which may be in your owner's manual or under the battery. Just a warning: one blogger at ZDnet.com wrote about his attempt to get the SAR rating for his phone, and it isn't always easy. Be prepared to be frustrated.

If you go to the FCC site, you'll see links to a variety of manufacturers from which you can get SAR numbers for your phone. If yours isn't there or you can't find it, you can also go to a second website (www.fcc.gov/oet/ea/fccid) where you'll find instructions for entering the FCC ID number to get the SAR from the FCC. Enter the number (in two parts as indicated: the Grantee Code is the first three characters, the Equipment Product Code is the remainder of the number). Then click on Search. You should see the grant of equipment authorization for your particular ID number. You'll probably find the highest SAR values reported in the equipment certification test data in the comments section of the grant of equipment certification.

What's the magic number? To pass FCC certification, a phone's maximum SAR level must be less than 1.6 watts per kilogram. This assumes about a six-minute exposure. In Europe, the level is 2 W/kg, while Canada allows a maximum of 1.6 W/kg.

Some electronics and computer magazines or websites will occasionally list the highest and lowest SARs of cell phones. But remember—those ratings apply to adults, not children who may absorb more of the radiation because of their smaller, thinner skulls and who may be at greater risk because their cells are still developing. And don't forget that safety ratings are based on the heating effects of radio frequency waves, meaning that the greater the heating effects, the more potentially harmful. Since studies have found biological effects at levels lower than official safety limits, including cellular DNA damage, paying attention to the SAR is a practical first step. More important is the time spent holding a cell phone, even with the best SAR, to your head. Invest in an air-tube headpiece, text as often as possible, and follow other safety tips below.

Put them on speaker. Anything you can do to keep the cell phone as far away from your head as possible will reduce the energy or power level because the farther away you are from the antenna, the lower the signal. (Even the FDA has recommended this, although in June 2009, it pulled down its web page on cell phone dangers and substituted one touting a new no-risk stance.) For example, if you hold it two inches away, the signal is about a quarter of its original strength. At four inches away, it's about one-sixteenth as strong. Whenever possible, use speaker phone mode or a hands-free kit with a wireless air tube nearest the earpiece, which is a type of headset available online and in many stores where phones are sold. (The wire on many headsets can act as an antenna which can deliver a dose of EMF radiation to your head.) Many of these air tube headsets sell for less than twenty dollars. You can also add a wire guard which is a clamp-on device that contains ferrite beads that suppress high frequency energy (see resources).

Use your words. Text whenever you can—it limits the duration of your exposure and keeps the phone farther away from your head and body. A phone with a keyboard rather than a typical phone pad makes texting much easier (though you may never get as good as most teens who say they can text blindfolded). When you text, don't keep the phone in your lap, however. There have been an increasing number of studies that have found damage to sperm vitality and motility in men who keep their cell phones in their pockets.[18]

Go offline. Make it a habit to turn the phone off when it's not in use or to switch it into offline, standalone, or flight modes, which turn off the wireless transmitter but still allow you to use the phone or PDA for everything except making and taking calls or browsing the web or e-mail.

Make the switch. If you absolutely must place the phone against your head (and I definitely do not recommend this) switch ears regularly while chatting on a cell to limit prolonged exposure on one side, which has been linked to increased risk of brain tumors and salivary cancers on the side of the head where the phone is usually held.

Avoid tight spaces. Don't make or take calls in the car—which thankfully is becoming increasingly against the law because it creates distractions—in elevators, trains, buses, or underground. First, your cell has to work harder to get a signal out through metal, so the power level increases. And on top of the more powerful signal, any metal box like the car or an elevator will also cause the waves to bounce around, creating in effect a resonance chamber, boosting their intensity.

Keep an eye on the bars. Don't use your phone when the signal is weak or when you're traveling at higher speeds in a car or train: this automatically boosts power to maximum as the phone attempts to connect to a new relay antenna.

Ride the quiet car. Many trains have so-called quiet cars where cell use is prohibited and phones must be switched off so they don't disturb other riders. It's your best bet for traveling without overwhelming secondhand exposure to RF radiation.

Keep it short. A cell phone isn't what you want to use to catch up with an old high school buddy. If your conversation is going to be long, use a landline. One study found that after two minutes, the brain's electrical activity can be altered for at least an hour.[19] If you're one of the roughly one in six Americans who are cell-only, think about getting a landline too, and use the cell only when necessary. Remember, brain tumor risk starts at a relatively low level of cumulative lifetime exposure.

Spend even less time on your PDA. Wireless devices such as the BlackBerry, iPhone, and Treo (palm-size devices known as personal digital assistants) produce higher ELF emissions than cell phones because they rely on the energy from batteries to power up things like e-mail and Internet connections and color displays, according to a study by EMF experts Olle Johansson, M.D., of the Karolinska Institute and Cindy Sage of Sage Associates. Both experts recommend never carrying the device in a pocket, keeping it at a distance in the office, and avoiding use when pregnant because some studies have shown an increased risk of miscarriage when a woman is exposed to 16 mG of ELF (the PDAs the researchers tested had emissions up to 975 mG).[20]

Dial, then stretch. Don't place the cell phone on your ear while your call is connecting—that's the time the phone is sending out its strongest signal.

Get it out of your pocket. A recent study found that men who carried their cells in their pockets had 25 percent lower sperm counts when compared to another group that didn't carry a cell.[21] Different parts of the body absorb radiation in different intensities, and testicular tissue may be more vulnerable. (That means no texting on your lap either!) Many experts also caution pregnant women about toting their phones because of unknown risks to a developing fetus. If you have to carry your phone with you all the time, tuck it in a purse or briefcase. There are also holsters made of shielded material that reflects the phone's radiation away from your body.

Keep the cell out of the bedroom. Specifically, don't sleep with your cell near your head. Remember, EMFs can reduce your body's production of melatonin and with it a powerful free radical scavenger that can protect your cells from the DNA damage that can lead to cancer and other disease.

CELL TOWER SAFETY

The cell tower in your neighborhood is a far tougher problem to solve, particularly if you're not able to move to a safer location. The best you can do is take steps to minimize your exposure.

For example, you can repaint the interior and exterior of your home with EMF shielding and conductive paints, products made with carbon, copper, silver, and nickel which can be applied like regular paint and then covered with a more aesthetically pleasing latex paint or wall paper. I did this in the master bedroom and home office. I can now sleep more deeply and focus much better!

It is very easy to measure the effects of paints, filters, and shielding fabrics with an RF meter. Your best bet is to make sure you diminish your exposure in ways that you can control, such as reducing emissions and your proximity to them inside the house, your car, and public transportation. In chapter 12 I will provide you with another tool, the Zapped Diet, which contains the nutrients your body needs to help neutralize the stressful impact from EMF side effects.

Waiting for Final Word

If there's one thing that demonstrates the chasm between scientists over the health effects of EMFs, it's the Interphone study. The twenty-four-million-dollar study (funded in part by the mobile phone industry) started in 2000 and combines the research on the link between cell phone use and brain cancer and other anomalies by scientists in thirteen countries—Australia, Canada, Denmark, Finland, France, Germany, Israel, Italy, Japan, New Zealand, Norway, Sweden, and the United Kingdom, all using the same protocols. There are a total of seventy-four hundred subjects, all of whom suffer from some kind of brain tumor, including glioma, meningioma, acoustic neuroma (a tumor of the acoustic nerve, generally nonmalignant), and cancer of the parotid gland. They weren't easy to find. These are all rare tumors, so rounding up more than seven thousand sufferers—especially since they all had to be cell phone users—was by no means an easy feat.[22]

Individual research groups have published studies and they have been, as others before them, contradictory. A pooled group of Scandinavian research has shown an increase in brain tumors and acoustic neuromas in heavy users of cell phones. (In the studies, heavy users and those who run the highest cancer risk are defined as those who have used a cell phone for more than forty-six months, spend more than five minutes on a call, and have spent more than two hundred sixty hours on the phone, which amounts to about twelve minutes a day.) Others have found no risks. Some teams, such as those from Canada and New Zealand, have chosen not to publish individual findings and are saving their results for the final paper. But the individual studies aren't considered as significant as the pooled data from all fifty scientists and their thousands of subjects.[23] After a "final" draft circulated the globe for four years while the researchers argued over what it should say, the Interphone study was released in May 2010. Some of the delay was caused when researchers felt the need to launch one last study to determine just how well cancer patients compared to a control group could recall how much they used the phone over the years. They found that everyone underestimated numbers of calls by 20 percent and overestimated the amount of time they spend on the phone by about 40 percent. This finding could lead to charges of "recall bias," which would cast the final report in doubt. Reportedly, some of the researchers have stopped talking to one another.[24]

What did the final report say? As expected, it was not conclusive. *Microwave News* had this to say when the results were released in 2010:

> *Everyone anticipated that Interphone wouldn't offer any definitive findings, and they were right.*
>
> *"An increased risk of brain cancer [has not been] established," said Christopher Wild, the director of the International Agency for Research on Cancer (IARC) in Lyon, which coordinated the study.*
>
> *However, there are "suggestions of an increased risk" at "the highest exposure levels," according to the abstract of the paper published by the* International Journal of Epidemiology.
>
> *How should those "suggestions" be interpreted?*
>
> *At the very least, the risks are greater than many believed only a few years ago. In a series of interviews, a number of the members of the Interphone project told* Microwave News *that they now see the risks among long-term users as being larger than when the study began. Some think the risks warrant serious attention.*
>
> *"To me, there's certainly smoke there," said Elisabeth Cardis, who leads the Interphone project. "Overall, my opinion is that the results show a real effect." Cardis is with the Center for Research in Environmental Epidemiology (CREAL) in Barcelona, where she moved two years ago after working on Interphone at IARC for close to a decade.*
>
> *Siegal Sadetzki, the Israeli member of Interphone who is with the Gertner Institute outside Tel-Aviv, pointed out that while the risks are inconclusive, a number of the results show some consistency. These include increased risks among the heaviest users, the fact that the risks were highest on the side of the head the phone was usually used and that the tumors were in the temporal lobe of the brain, which is closest to the ear.*
>
> *"The data are not strong enough for a causal interpretation, but they are sufficient to support precautionary policies," she said.*
>
> *One strong dissenting voice is that of Interphone's Maria Feychting of the Karolinska Institute in Stockholm. "The use of mobile phones for over ten years shows*

no increased risk of brain tumors," stated a press release issued by the Karolinska. Feychting declined to be interviewed for the Microwave News article."[25]

Safer Phones

In 2010, the Environmental Working Group (EWG) published its list of cell phones and PDAs from major carriers emitting the lowest amount of radiation. Here are their choices:

1. Sanyo Katana II, Carrier: Kajeet
2. Samsung Rugby (SGH-a837), Carrier: AT&T
3. BlackBerry Storm 9530, Carrier: Verizon Wireless
4. Samsung I8000 Omnia II, Carrier: Verizon Wireless
5. Samsung Propel Pro (SGH-i627), Carrier: AT&T
6. Samsung SGH-t229, Carrier: T-Mobile
7. Helio Pantech Ocean, Carrier: Virgin Mobile
8. Sony Ericsson W518a Walkman, Carrier: AT&T
9. Samsung SGH-a137, Carrier: AT&T
10. LG Shine II, Carrier: AT&T

Safer Smartphones (PDAs)

1. Nokia 7710
2. BlackBerry Storm 9500 Smartphone
3. BlackBerry Storm 9530, Carrier: Verizon Wireless
4. Nokia 9300i
5. Samsung I8000 Omnia II, Carrier: Verizon Wireless
6. Samsung Propel Pro (SGH-i627), Carrier: AT&T
7. T-Mobile Wing (HERA110), Carrier: T-Mobile
8. Nokia E90
9. Helio Pantech Ocean, Carrier: Virgin Mobile
10. Samsung Flight (SGH-A797), Carrier: AT&T

You can see where your phone falls on the list by going to the EWG website at www.ewg.org. The group ranks more than one thousand cell phones on the market.

9:

ZAP-PROOF
YOUR KIDS

As you might expect, protecting children from electropollution begins even before a child is born. And so, in this chapter you will be introduced to many tips and tools you need to safely raise a healthy family today. Just keep in mind that infants and children are not just little versions of you and me. Instead, their bodies are significantly different than adult bodies and also far more susceptible to potential damage. The bottom line: yes, children are definitely at risk and you may never have suspected the extent to which they are being exposed to electropollution.

So you may be startled at what you will read in this chapter. But you have the power to make positive changes in the lives of your children, and you can take charge and make a lasting impact in their long-term well-being. The gift of your children's future health is in your hands today.

Four-year-olds with working cell phones? It's happening. Parents who wouldn't dream of letting their kids use a microwave oven are buying their kids phones that will expose them to a similar form of EMF—this one that can penetrate across the depth and breadth of a child's brain.

Miami-based Firefly Mobile markets a phone that's tiny enough to fit easily into a preschooler's hands, though the company says it makes the phone for kids of all ages. Even Disney produces a phone for tweens and teens (kids eleven through fifteen) that has plenty of flashy bells and whistles—camcorder, camera, games, and customizable ring tones—to keep kids on the phone so long they'd have to be classified as heavy users even if they never make or receive a call.

Parents love cell phones because it allows them to reach their kids—and to be reached—in case of an emergency. And that's a good thing. But they may be inadvertently putting their children at another risk.

WHY CHILDREN ARE MORE VULNERABLE

Children are always the most vulnerable victims of environmental pollution. For instance, because of their size, they absorb 50 percent more air pollution than adults.[1] When it comes to EMFs, studies have shown that their thinner skulls and bones allow them to absorb twice the amount of radiation as a grown-up. EMFs can more deeply penetrate their brain tissue, which is more conductive than an adult's because it contains a higher concentration of fluid and ions.[2] One study found that a cell phone call lasting only two minutes causes brain hyperactivity that persists for an hour in children.[3]

Because they're growing 24/7, children's cells are dividing at breakneck speed. The more cells that divide, the greater the risk for damage at critical junctures. EMFs can cause breaks in the blood-brain barrier, which is more permeable in children.[4] That breach can allow toxins to travel to the brain through the bloodstream and can result in oxidative stress, damage to nerve tissue, and adverse effects to brain hormones such as dopamine and serotonin.

Worse, children may be exposed, even before they're born, to the pollution a mother *absorbs*—through her cell phone use, even her exposure to high EMFs at home or on the job. At a 2004 World Health Organization meeting on EMFs and children, experts pointed out that RF emissions can cause whole body heating.[5] Ultrasound scans, now done routinely on women during pregnancy, produce strong frequencies and penetrate the body deeply, creating heat in the tissues of both the mother and baby. Heat—when it raises the mother's core temperature to 102 degrees—can result in congenital abnormalities, and the effects are cumulative. Certain tissues in the body absorb frequencies more readily; it all depends on their fluid content—the more fluid they contain, the more EMFs are absorbed. Since a fetus spends forty weeks essentially floating in a sack of water, it is likely that it's particularly vulnerable.[6]

Many studies have linked maternal and child exposure to EMFs, even at levels regulatory agencies consider safe, to childhood cancers (particularly leukemia) as well as asthma, autism, ADD, and behavioral problems in children, rates of which have been rising at an alarming rate. For example, alternative physician Dietrich Klinghardt, M.D., in a small but significant study, found that a pregnant woman's body voltage, measured in her bedroom, as well as the body voltage of children in their bedrooms, was a predictor of autism and other serious neurological disorders.[7]

As a parent, you want what's best for your child. Before your baby is born, you scan the house for potential dangers and baby-proof it with locks and outlet covers. Think of this chapter as an instruction booklet for EMF child-proofing your kids' lives, from the womb until they leave the nest.

FROM THE WOMB TO ADULTHOOD: ELECTRONIC SAFETY FOR YOUR KIDS

Today, an estimated thirty-one million kids—10.5 million of them preteens—are on their cell phones on average 3.75 hours a day. That's a lot of close exposure to RF waves. Here's how to help them cut way back, starting in the womb!

First and foremost, restrict cell and cordless phone use during pregnancy. Here's why: A 2008 survey of more than thirteen thousand children found that women who use a cell phone two or three times a day while pregnant are more likely to have children with behavior problems such as hyperactivity, difficulty controlling their emotions, and developing good relationships with others. In the same way that you protect your baby from other toxic exposure during pregnancy, it is vital that you protect your baby from electropollution, as it can impact the baby's nervous system development—long before birth.[8] Perhaps more important, heavy-duty cell phone use during pregnancy has been linked to risk of miscarriage and birth defects.[9]

Are Teens Addicted to Cell Phones?

Many teens think they are addicted to cell phones, and they certainly act like it. In one study, teens aged fifteen to nineteen describe themselves as addicted to their phones, and in another, 30 percent say they're depressed when they can't use it.[10]

Two Latin American researchers—a psychologist who works with drug addicts and an EMF expert—in a paper published in March 2007 suggest a plausible explanation for why teens go into withdrawal when they can't use their phones: cell phone use, like drugs and alcohol, may act on the reward centers of the brain which contain opiate receptors. When the brain gets its perceived reward—whether it's heroin, chocolate, or the fun of texting two thousand times a month on average—it wants a do-over, again and again and again. Suddenly, a behavior is born.[11]

In fact, they note, Henry Lai, Ph.D., of the University of Washington has found that microwaves increase the activity of the feel-good hormones known as endorphins and the brain's own happy chemicals—neurotransmitters such as dopamine. Lai further ties EMFs to addiction: using a drug that quashes heroin cravings, he blocked these effects in rats exposed to RF waves.[12]

Psychologists who have studied cell phone use, particularly texting, by teens, report that it's leading to anxiety, behavioral problems, distraction in school, repetitive stress injury, and sleep deprivation.

If your teen is addicted, he's not going to stop on his own. You may need to enforce the age-old rule that the one who pays the bill (at fifteen cents apiece for texts, likely no small figure) gets to decide how and how often a phone is used.

Don't throw away your cell phone! But use it only for emergencies, not casual chatting or lengthy conversations. Text when you can so you avoid holding the phone near your head. If you must use a cell phone, buy a hands-free kit with a wireless air tube near the earpiece. Carry your phone in a purse rather than in a pocket or clipped to a belt—unless it's turned off, a cell phone is always emitting dangerous microwave radiation which can affect you and your unborn child. Follow the other tips on safe cell phone use in chapter 8.

Minimize total prenatal AC magnetic field exposure. Before you get pregnant, do a home inspection with a Gauss meter that picks up AC magnetic field readings. Ideally, you want your home's AC magnetic field level to be no more than 1 mG if there are children or if you're planning to start a family. If you find higher readings, find the spot where they begin to fade and position furniture that safe distance away from the EMF source.

Reduce dirty electricity. Measure dirty electricity with the Stetzerizer Microsurge meter at various outlets throughout your home. If readings are very high, you will want to consider installing the special filters, called Graham-Stetzer filters, that are discussed in chapter 7. These, as you may remember, neutralize the medium frequency voltage spikes (2,000 to 100,000 Hz), a peculiar and dangerous hybrid of low and high frequency waves known as dirty electricity. Follow the instructions in chapters 5, 6, and 7 for inspecting your home and taking measures to eliminate any dangerous EMF emissions from the environment.

Sweep the nursery and kids' bedrooms. Avoid baby monitors if you can, but if you need one for your peace of mind, look for a used older monitor that is a wired baby monitor. Newer monitors with the stubby half-inch antenna operate the same as a Wi-Fi or cordless phone and should not *ever* be used in the house, let alone in a bedroom. Keep them as well as night-lights or bedside lights six feet away from your child as he sleeps. Determine where wiring is in the wall, and check beside and below the child's bedroom for water pipes, appliances, computers, garage door openers, black box transformers, and fluorescent lights—all sources of low and medium frequency AC magnetic fields—and position your baby's crib or child's bed as far away from them as possible. Use a Trifield or Gauss meter to measure precisely: you want to keep children as far away as possible from AC magnetic fields measuring more than 1 mG. Measure RFs with an electrosmog meter.

Microwave food without baby. Baby's cranky time usually coincides with dinner preparation, but resist the urge to carry your crying child on your hip while cooking—especially at the microwave. The kitchen is filled with EMF-producing appliances, the most powerful of which is your microwave.

Curb kids' cell phone use. In Russia, scientists and government officials have advised that anyone under the age of eighteen should not use a cell phone. In France, there's a ban against marketing mobile phones to young children. In the United Kingdom, Sir William Stewart, chairman of the National Radiological Protection Board, was quoted as saying, "I don't think we can put our hands on our hearts and say mobile phones are safe."[13] In the United States? Not much . . . yet.

In an ideal world, you would never buy your child a cell phone. But if you did that, you might never hear the end of it because all their friends now own them. Besides, in our two-career families, those cell phones equate to peace of mind. But since you're the parent, you can set the rules for cell phone use.

First, do not buy preschoolers cell phones. It is far too dangerous a device to put into their hands—and especially for them to put up to their ears.

There are studies suggesting that, in addition to brain tumors, children's cell phone use could result in hearing loss (if they use it an hour a day or more) and the death of auditory nerve cells. One Indian study found that cell phone users who were on the phone for sixty minutes a day over four years experienced damage to the DNA in roughly 40 percent of their cells.[14]

If you absolutely must equip your little one, look for phones without dial pads. Those will have buttons that *you* can program to speed dial emergency numbers only. And insist they be carried in shielded cases.

Cell phones are just as dangerous for your tweens and teens as they are for your preschoolers. But now, peer pressure *really* kicks in. As long as parents are paying for cell use, however, you have the power of the purse. Parents of both preteens and adolescents will need to establish clear guidelines as to their cell phone use, while understanding your kids' needs for independence. Your children will be even safer if they obey these rules:

- Text rather than call.

- Use a safe headset.

- No phoning or texting while driving. The U.S. Department of Transportation reports that teen drivers on cell phones have slower reaction times than teens not on the phone while behind the wheel.[15] British scientist Andrew Goldsworthy has suggested that it's not just inattention that triggers car accidents involving cell phone use, but the false signals our nerve cells send out when they leak, as they do when exposed to EMFs. The brain hyperactivity that results may also be at the root of attention deficit hyperactivity disorder (ADHD) in children, he speculates.[16]

- No phones in the bedroom. Many teens already have sleep problems, and a constantly beeping or buzzing phone—and friends just a text message away—won't help. (While you're at it, move all the electronics from sleeping areas to avoid EMF exposures when kids are most vulnerable.)

- Teach kids to keep phones away from the body while they're turned on. Those incoming calls and texts cause a burst of RF emissions that can penetrate the body, so have them stash the phone in a purse or backpack.

As a parent, the more you can do to set a good example with your own cell phone use—making calls on a landline at home, for example— the more likely your kids are to follow suit. Don't be afraid to ground any child's cell phone use—by taking it away for a day or two if she ignores reasonable rules!

Fight Wi-Fi in schools, communities—and at home. It's easy enough to stick with or go back to wired technologies at home. It's a little harder to protect your family when your school district or city wants everyone wireless. While schools and even cities now consider Wi-Fi de rigueur, many parents, thankfully, do not. In one suburb of Chicago, concerned parents filed suit against the Oak Park School District, claiming that constant exposure to 802.11 Hz signals could cause their children neurological harm. (They cited one Swedish study that showed that exposure caused brain damage in teenaged rats.[17]) In England, consumer, parent, and teacher groups have all called for a suspension of Wi-Fi in schools.[18] In Santa Fe, New Mexico, a college librarian quit her job because she didn't think that "allowing oneself to be irradiated" by Wi-Fi should be a condition of employment.[19] If your school or community is considering going wireless, organize with other concerned people and use the information in this book to campaign against it.

Campaign against cell towers and antennas. For all you know, that church steeple or grain silo near your child's school hides a cell phone tower or antenna. Or there might be one on the roof of the school itself. Go to www.antennasearch.com to find any cell towers or antennas near your child's school. Studies have found an increase in cancer in people who live within about one-quarter of a mile of towers and antennas, so if they are this close to your child's school, work with other parents to convince your school or school district to request that local authorities have the towers and antennas moved.

You may run into opposition, but maybe not. In September 2009, school officials made the unprecedented move to close the Fredon Township Elementary School in New Jersey, because AC magnetic field readings from the eighty-foot electrical towers carrying 230 kV lines across school property were six times the recommended safe levels set by the World Health Organization (which is still higher than most EMF experts recommend).[20]

The community became concerned when the local utility proposed an upgrade to the electrical system, a project that would string 500 kV transmission lines from eastern Pennsylvania to New Jersey. For the first time, readings were taken on the existing lines and the results were shocking: AC magnetic field levels averaged 19.34 mG. The WHO recommends a 3 mG threshold, while other experts recommend 1 to 2 mG.[21]

Fix your electrical system. To its credit, the power industry has not dismissed the possible connection between electromagnetic fields and childhood leukemia, a link first made in 1979 when researchers Nancy Wertheimer and Ed Leeper found that children who lived closest to utility power lines that could be expected to produce elevated magnetic fields in the home transformers had a two- to three-fold increase in cancer, particularly leukemia and brain tumors. Since 1999, the Electric Power Research Institute (EPRI) has been pursuing a plausible explanation for the leukemia findings confirmed by international studies.[22]

Since 1999, EPRI is following a promising lead: contact current. That's the current that flows within you when you're in contact with two electrically conductive surfaces with different voltages, like when you place your thumb and forefinger on either end (positive and negative) of a battery. You may not feel it, but a current flows through you. It's likely to be fairly small unless your fingers are wet, which would give you an electrical shock since water is a much better electrical conductor than skin. Don't try this.

Where does EPRI think children are becoming the conduit for an electrical current in the home? In the bathroom.[23]

As I've pointed out, since 1918, the National Electrical Code has required that your electrical system be grounded to one metal water pipe, which gives the pipe what's usually a less than 1 mV charge as the electric current travels along the pipe back to the nearest power substation. In chapter 6, I showed you that the journey isn't quite as straightforward as it sounds: current can stray far and wide and pick up unwelcome high frequency hitchhikers as it heads home. It can also travel throughout your house along metal pipes (only metal pipes are conductive), even creating a current in older ceramic-coated metal bathtubs or sinks (or your stainless steel refrigerator with ice maker).

If a child touches a faucet, for example, that current will travel into her arm and throughout her body—even into her bone marrow! A child, especially one who is wet, can expect to get the biggest dose in the thinnest extremities, such as a hand or arm. A study done at the University of Victoria in British Columbia, Canada, confirmed that contact current so mild it was imperceptible could produce a magnetic field in the bone marrow of the forearm as much as thousands of times higher than the normal environmental field.[24]

EPRI researchers knew that children were very likely to touch and play with faucets while in the tub or at the sink. They worked with scientists at the University of California at Berkeley who, using interviews and diaries, found that 80 percent of the children from infants to five-year-olds whose parents reported on their tub activities frequently touched the faucet and played in the water stream.

The research is ongoing. EPRI has teamed with the long-running Northern California Childhood Leukemia Study to determine if contact current in homes is associated with an increased risk of this childhood cancer. But you don't have to wait for an answer. See chapters 5 and 6 for ways you can test for and reduce contact current in the home.

THIS IS YOUR CHILD'S BRAIN ON EMFS

Rates of autism, ADHD, and learning disabilities have soared in the last forty years. In 1970, for example, about one in ten thousand children was diagnosed with autism. Today, the incidence is estimated at one in one hundred and ten.[25] No one is sure why. Genes, greater awareness, and better diagnostic techniques may all play a role. But is it really a coincidence that these statistics have gone up at the same time as burgeoning technologies are surrounding us with EMFs?

We know that EMFs cause a disruption in cellular communication, in part by causing membrane rupture and leakage of calcium ions, which play a major role in transmitting messages throughout the body. Andrew Goldsworthy, who has written extensively on the subject, believes that when sensory cells leak their calcium, they can "send false signals to the brain," which can result in sensations like pins and needles or burning.

When neurons—nerve cells—leak, they also start transmitting impulses, which Goldsworthy believes can cause hyperactivity in the brain and may be at the bottom of at least some cases of ADHD.[26] Early animal studies did find that exposure to an everyday 60 Hz field interferes with the ability of rats and monkeys to perform simple tasks, including pressing a lever for food.[27]

Other researchers found that the kinds of lesions identified in the brains of some children with autism are the same as those that were created in animals exposed to 50 Hz fields (the same as the electrical system in Europe) and to microwaves. In a study published in 2006 in the journal *Medical Hypotheses,* researcher I. M. Thornton of the University of Wales postulated that environmental exposure to EMF could affect mirror neurons, the nerve cells that fire when we observe an action in another or act ourselves. (Those neurons allow use to take behavior we've seen and turn it into behavior we do.) At least one theory of autism is that the mirror neuron system is faulty, throwing a child's perception off and making him unable to mimic the actions of others or to act normally himself.[28]

Thornton speculates that EMFs could interfere at a critical point in the maturation of a child's brain when the neurons of this system are grouping to work together. In adults, this see-and-do system is already formed.[29]

Tamara J. Mariea, CCN, who has treated more than five hundred children with autism spectrum disorder at her Tennessee Clinic since 2000, published a study in 2007 in the *Journal of the Australasian College of Nutritional and Environmental Medicine* reporting her success in improving her patients' symptoms without changing any of their previous treatment. The only thing she did differently was lower the EMFs in her clinic.[30]

Remember, you can't rely on federal or industry safety standards when it comes to kids' exposure to EMF. Not only are those rules governing communications radiation only concerned with thermal effects—whether you can be burned by the radiation—they are based on the height and weight of a six-foot man, not a teen, child, or baby.

Now that you are off to a running start with the kids, let's take a moment and revisit—but in more detail—the specific zapped challenges in the workplace and what you can do to eliminate them. This next chapter is a must for anyone who works, especially for managers, owners, and corporate executives.

10:

ZAP-PROOF YOUR WORK ENVIRONMENT

I n retrospect, Gilligan Joy picked the wrong career.

When Gilligan, a native of British Columbia, Canada, was studying electronics in college, he found himself becoming easily fatigued. But it wasn't until he was finishing up a bachelor's degree in computer science in the 1990s that his tiredness became so overwhelming that he had to take frequent naps every day.

With degree in hand, he got a great job as a software developer for the Canadian government at the naval base in Esquilmalt, British Columbia, a base with a radar station and ships equipped with radar antennas. Pretty soon, Gilligan was taking naps on his lunch hour and breaks to replenish his dwindling strength. He had to rest the minute he got home.

"It got to the point that when a co-worker would turn on the fluorescent lights, it would drain my energy level," he recalls. "I figured it had to be something at work because I would partially recover on the weekend."

He thought that he might have been reacting to light, so he experimented with working with the lights off, his office lit only by his computer. "My health improved dramatically," he says. "Fortunately, my co-worker who shared an office with me agreed to having the lights turned off. Thankfully—I couldn't have survived without his help."

Then, he got a clear, anti-radiation shield for his and his co-worker's computer, which was the older type with a high EMF cathode ray tube. "My health improved again," Gilligan recalls. His fatigue lifted, and he wasn't getting as many colds as he had before.

But in 1999, his health took a nosedive. Along with weakness, he now had nausea, abdominal pain, headache, gastric distress, a pounding heart, light sensitivity, sinus troubles, difficulty focusing, and fuzzy thinking. He nearly died.

He'd already done the reading—he knew he had become sensitized to the radar. His symptoms were those described in the medical literature as characteristic of radio wave sickness, a common complaint of World War II radar operators. He requested a transfer based on his medical condition, which was refused because the Canadian health program didn't recognize electrosensitivity as a real illness. However, his supervisors permitted him to work at home four days a week, which allowed him to recover somewhat but not totally.

He was denied disability and tried other kinds of work, but wound up back with the Navy where he hit upon a creative solution—hot water heater insulation with an aluminum cover, which he used as wallpaper in his office to shield him from the fields he believed were making him sick. Ultimately, though, with the help of his union, he was placed in a shielded room.

Gilligan was still working for the Navy when we spoke in early 2009—and out of the shielded room, thanks to a combination of alternative therapies such as daily breath work and a diet designed to restore health to his central nervous system. "I'm not free from sensitivity, but I'm close to being free," he says.

Gilligan Joy was remarkably lucky. Certainly not that he became electrosensitive, but that he found help in the workplace that allowed him to continue supporting himself with the skills that put him in harm's way.

Even if you're not hypersensitive to EMFs and don't work where you're bathed in radar waves, you may still be exposed to intense EMFs on the job that may be affecting you without your knowledge. Many of the symptoms we write off to stress—fatigue, headaches, aches, and pains—

are also signs of this modern, everyday version of radio wave sickness. If your desk is located near a transformer, an electrical closet, a circuit box, power cables, or other electrical equipment, including factory machinery, you're exposed for about a third of your day—possibly more—to AC magnetic fields.

Like your home, your workplace may have a variety of magnetic field levels from very low to extremely high. Clearly, some occupations make you more vulnerable than others, a fact not lost on the hundreds of researchers who are investigating occupational exposures and risk of everything from breast cancer in men to ALS, or Lou Gehrig's disease.[1] For example, some studies have shown that American electrical utility workers with the most AC electric and magnetic exposure are twice as likely to die of prostate cancer as those with the lowest exposure.[2] One study looking at data on more than one hundred thirty thousand electrical workers found a link to increased risk of suicide.[3] Other studies have found an association between EMF exposure and depression and other psychiatric illness.[4]

A University of Washington study found that electrical and communication workers had higher rates of breast cancer—and this study was only looking at men.[5] Other studies have pointed to those in the welding and engineering professions too.

The breast cancer link was first noted in a study published in 1991. It found that rats exposed to EMFs had a much higher incidence of breast tumors than those who weren't exposed. A Norwegian study found that men were getting breast cancer at twice the expected rate when exposed to EMFs, which launched an initiative by the Occupational Safety and Health Administration (OSHA) to investigate the link.[6]

Breast cancer is highly unusual in men. In fact, it is one hundred times less common than it is in women. Fewer than two thousand cases are diagnosed each year in the United States, and about four hundred men die annually of the disease, which kills about forty-six thousand women annually. According to the National Cancer Institute, the top risk factors for male breast cancer are family history, having a disease that raises estrogen levels, and exposure to radiation.[7]

I know what you're thinking now. What am I supposed to do, quit my job? It may not be necessary to take drastic measures when your problems can be greatly diminished by making a few simple—or one or two major—changes. It could be as easy as moving your desk a couple feet.

Here are a few things that can make an impact.

Bring your Gauss and electrosmog meter to work. Take measurements at your work area. Pay special attention to readings near where you sit or stand. You may be surrounded by office equipment emanating high fields, but if your usual spot is far enough away, you may be less exposed than you think.

Survey the territory. Identify any areas that may be more exposed than others, such as locations near electrical utility closets, electrical boxes, cables, transformers outside the building, cell or radio frequency towers (remember, they could be on the roof of your building or nearby and may be disguised).

Elevators may also be a concern. At the University of California at San Diego, a higher than normal number of employees working in the literature department were diagnosed with cancer. Epidemiologist Cedric Garland, M.D., from the university's Department of Family and Preventive Medicine, conducted a study which concluded that women who worked in the building had a chance of developing breast cancer that was four to five times greater than if they didn't work there. Other potential causes, such as radioactive chemicals and carcinogens, were ruled out.[8]

Garland suggested the link could be the building's hydraulic elevator, which produces a surge of power every fifteen to sixty seconds with a corresponding spike in AC magnetic field far above safe limits. His findings were controversial—of course—and a second study was under way as of this writing.[9]

Adjust your position. If you can, move your desk chair away from the wall where wiring is located or where magnetic field–generating machines—such as copy machines—are on the other side of the wall. Keep wires, surge protectors, extension cords, and power bars at a distance too—six feet is usually recommended, but at least three will diminish the fields they generate. When you're making copies, don't hover over the copy machine, which may produce EMFs of varying frequencies. Step back a few feet while it's in operation. In many offices, computer hard drives aren't on the desktop but under the desk. Move them as far away from your body as you can. Don't create a horseshoe of equipment around you—that will expose you to the EMFs from many machines at one time. And turn off all equipment when you're not using it.

Talk to your supervisor. If your office, workspace, or classroom is located near a source of high EMFs, talk to your boss about moving. You may want to show him or her the readings your Gauss meter shows around your desk or work area. Some bosses may ask to borrow your meter, others may be unconcerned. Use discretion. You may also want to give them a copy of this book to peruse. Your best bet is to recommend simple, free, or low-cost solutions to any EMF problems in the workplace. Keep emotions out of it (don't say, "This stuff could be killing me!" or indulge in nonprofessional histrionics) and talk about the scientific evidence. Anyone who runs a business is all about the statistics and data—use them. Health care costs eat up a lot of revenues—your boss may be happy to do anything that will reduce that particular business expense.

Bring in your utility. If you work in an industry where you're around large motors, you could be needlessly exposed to high EMFs because the right filters haven't been installed. If you work with or around variable-speed drives—you find them in any type of equipment where the speed needs to be adjustable, for example in ventilation fans, assembly lines, pumps, and agricultural machinery, such as milking machines—you may be bathing every day in fields that fall into the category of dirty electricity. A few thirty-five-dollar harmonic noise filters can quash many of them. The benefit for your employer is extending the lifetime of an expensive machine and reducing both equipment and employee downtime.

Talk to your union. Gilligan Joy's union helped him convince his supervisors that he needed to work in a shielded office. Unions were started to help protect the health and safety of workers—and they still follow that mission.

Contact NIOSH. The National Institute for Occupational Safety and Health sometimes conducts health hazard evaluations in workplaces where EMF may be a serious concern.

Consider filters. Schools may be hotbeds of dirty electricity, which has been shown to cause fatigue, chronic illness, behavior problems, headaches, asthma and allergy attacks, and learning difficulties in students. Teachers are also affected. In one school in the Desert Sands Unified School District, in California, researchers found a cancer cluster among its

137 teachers—eighteen cancers in sixteen teachers, which carries only a one in ten thousand possibility of being due to chance. The researchers, who were brought in by the teachers, not the school district, found that thirteen rooms in the school had dirty electricity, some at levels so high that they could not be measured by a state-of-the-art microsurge meter.[10]

Studies by Magda Havas, Ph.D., a professor of environmental studies at Canada's Trent University, have found that Graham-Stetzer filters can reduce all of those problems. In one study, for example, Havas found that children's use of asthma nebulizers dropped dramatically after the school installed the filters. Of thirty-seven students who previously used the treatment daily, only three used their nebulizers after the filters were in, and then only for exercise-induced asthma.[11] Havas also worked in several schools in Minnesota that looked at teacher health, including both physical and emotional health: after the filters were installed, 64 percent said their health improved.

Switch out the lighting. Many office buildings, schools, and factories are equipped with horrible fluorescent lighting that produces higher EMFs than incandescent lighting. It's unlikely your company is going to replace all its light fixtures, but you may be able to convince managers to provide employees (at least those with desks, offices, or cubicles) with incandescent lamps that they can use instead of flicking on overhead lights. Some key selling features: Overhead lights, particularly in older buildings, often pour illumination into areas where it's not needed, which costs more in energy bills. (Newer designs place track lighting only over areas where light is needed.) Also, studies have found that workers find fluorescent light with its unnaturally green and yellow tint to be irritating, more frequently complaining about headaches, stress, and fatigue, which affect their work.[12] This kind of lighting can also be harmful to those with epilepsy (it initiates seizures) and migraine. Use incandescent and energy-efficient LED instead.

Use your cell phone and PDA safely. If you're in sales or you travel a great deal for business, you probably rely heavily on your cell to get business done. Follow the advice in chapter 8, and especially, do not carry your switched-on cell in your pocket. A study released in 2008 confirmed that cell phones could reduce the vitality and motility of sperm—keys to fertility.[13] Exposure also caused DNA damage.

Farmers, pay attention to your cows. In some places—in the Midwest especially—healthy farm living has become a myth. Utility companies use the earth as a return for neutral current spawning what's called stray voltage or neutral-to-ground voltage, which produces electrical shocks that can cause cows to reduce their milk production, get sick, or even die. Farmers, too, have been affected (see Catherine Kleiber's story in chapter 6). Cows tend to be more sensitive to small voltages than people, so watch for this kind of behavior in your animals:

- Cows are hesitant to step into the milking parlor, barn, or other places on the farm. They may make wide circles around an area, or stampede out of a building.

- Animals are reluctant to eat or drink.

- Cows suddenly twitch or act nervous for no reason.

- Normal milk production is diminished.

If you see any of this behavior, call your local utility and ask a representative to come out and take measurements, particularly in the areas where you suspect a problem. Stray voltage can also be the result of poor or faulty wiring, faulty equipment, and improper grounding, so call an electrician too.[14]

Ask for a Wi-Fi–free zone. See if your employer will provide you with wired Internet capabilities.

Next up, find out how to protect yourself from the other zappers in your life.

EMFS OF SOME TYPICAL OFFICE EQUIPMENT

Magnetic field measurements are in units of milliGauss (mG)

	Distance from Source			
	6 inches	1 foot	2 feet	4 feet
Air Cleaners				
Lowest	110	20	3	—
Median	180	35	5	1
Highest	260	50	8	2
Copy Machines				
Lowest	4	2	1	—
Median	90	20	7	1
Highest	200	40	13	4
Fax Machines				
Lowest	4	—	—	—
Median	6	—	—	—
Highest	9	2	—	—
Fluorescent Lights				
Lowest	20	—	—	—
Median	40	6	2	—
Highest	100	30	8	4
Electric Pencil Sharpeners				
Lowest	20	8	5	—
Median	200	70	20	2
Highest	300	90	30	30

Source: "EMF in Your Environment," Environmental Protection Agency (EPA), 1992.

AVERAGE DAILY EMF EXPOSURES FOR CERTAIN OCCUPATIONS

EMF magnetic fields are measured in milligauss (mG)

Industry and Occupation	Median for Occupation	Range for 90% of workers
Electrical Workers		
Electrical engineers	1.7	0.5–12.0
Construction electricians	3.1	1.6–12.1
TV repairers	4.3	0.6–8.6
Welders	9.5	1.4–66.1
Electric Utilities		
Clerical workers without computers	0.5	0.2–2.0
Clerical workers with computers	1.2	0.5–4.5
Line workers	2.5	0.5–34.8
Electricians	5.4	0.8–34.0
Distribution substation operators	7.2	1.1–36.2
Workers off the job	0.9	0.3–3.7
Telecommunications		
Install, maintenance, and repair technicians	1.5	0.7–3.2
Central office technicians	2.1	0.5–8.2
Cable splicers	3.2	0.7–15.0
Auto Transmission Manufacture		
Assemblers	0.7	0.2–4.9
Machinists	1.9	0.6–27.6
Hospitals		
Nurses	1.1	0.5–2.1
X-ray technicians	1.5	1.0–2.2

Industry and Occupation	Median for Occupation	Range for 90% of workers
Selected occupations from all economic sectors		
Construction machine operators	0.5	0.1–1.2
Motor vehicle drivers	1.1	0.4–2.7
School teachers	1.3	0.6–3.2
Auto mechanics	2.3	0.6–8.7
Retail sales	2.3	1.0–5.5
Sheet metal workers	3.9	0.3–48.4
Sewing machine operators	6.8	0.9–32
Forestry and logging jobs	7.6	0.6–95.5

Source: National Institute for Occupational Safety and Health.

EMF SPOT MEASUREMENTS FOR INDUSTRY

Industry and Sources	ELF Magnetic Fields (mG)	Other Frequencies	Comments
Electrical equipment used in machine manufacturing			
Electric resistance heaters	6,000–14,000	Very Low Frequency (VLF)	
Induction heaters	10–460	High VLF	
Handheld grinders	3,000		Tool exposures measured at operator's chest
Grinders	110		Tool exposures measured at operator's chest
Lathes, drill presses, etc.	1–4		Tool exposures measured at operator's chest
Aluminum Refining			
Aluminum pot rooms	3.4–30	Very high static field	Highly rectified DC current with an ELD ripple refines aluminum
Rectification rooms	300–3,300	High static field	

Industry and Sources	ELF Magnetic Fields	Other Frequencies	Comments
Steel Foundry			
Ladle refineries			
Furnaces active	170–1,300	High ULF from ladle's stirrer	Highest ELF field was at the chair of the control room operator
Furnaces inactive	0.6–3.7	High ULD from ladle's stirrer	
Electrogalvanizing units	2–1,100	High VLF	
Television Broadcasting			
Video cameras (studio and minicams)	7.2–24.0	VLF	
Video tape degaussers	160–3,300		Measured 1 foot away
Light control centers	10–300		Walk-through survey
Studios and newsrooms	2–5		Walk-through survey
Hospitals			
Intensive care units	0.1–220	VLF	Measured at nurse's chest
Post-anesthesia care units	0.1–24	VLF	
Magnetic resonance imaging (MRI)	0.5–280	Very high static fields, VLF, and RD	Measured at technician's work location

Industry and Sources	ELF Magnetic Fields	Other Frequencies	Comments
Transportation			
Cars, minivans, and trucks	0.1–125	Most frequencies less than 60 Herz	Steel-belted tires are principal ELF source for gas/diesel vehicles
Buses (diesel powered)	0.5–146	Most frequencies less than 60 Hz	
Electric cars	0.1–81	Some elevated static fields	
Chargers for electric cars	4–63		Measured 2 ft. from charger
Electric buses	0.1–88		Measured at waist. Fields at ankles 2–5 times higher
Electric train passenger cars	0.1–330	25 and 60 Hz on US trains	Measured at waist. Fields at ankles 2–5 times higher
Airliner	0.8–24.2	400 Hz	Measured at waist
Government Offices			
Desk work locations	0–1.7		Peaks due to laser printers
Power center	18–50		
Power cables in floor	15–170		
Building power supplies	25–1,800		

Industry and Sources	ELF Magnetic Fields	Other Frequencies	Comments
Can openers	3,000		Appliance fields measured 6 inches away.
Desktop cooling fans	1,000		Appliance fields measured 6 inches away.
Other office appliances	10–200		

ULF (ultra low frequency) = frequencies above 0, below 3 Hz
ELF (extremely low frequency) = frequencies 3–3,000 Hz
VLF (very low frequency) = frequencies 3,000–30,000 Hz

Source: National Institute for Occupational Safety and Health, 2001.

Press for minimizing RF exposures that would include external sources (such as neighborhood antennas, towers, and antennas attached to buildings) and internal exposure for wireless equipment and secondary cell phone exposure.

OTHER ZAPPERS IN YOUR LIFE

I t all started with a persistent cough that I have had for twenty years—on and off. I had a gastroscopy to assess my stomach, esophagus, and duodenum, and nothing showed up. So I decided to do a total body scan to see what else might be going on, and sure enough, the technician noted a cyst on the right lobe of my thyroid. My integrative physician suggested a chest X-ray to rule out trachea displacement due to an enlarged thyroid, and I thought this was a good idea. Although I had concerns about the ionizing radiation from a chest X-ray, I felt this was essential to the resolution of my cough. Being a public speaker and doing ten interviews a week as well as at least twenty weekly consultations, the constant coughing fits were beginning to worry me after all these years.

I eagerly took the chest X-ray and had it evaluated by an ear, nose, and throat doctor. When the doctor looked at the X-ray and read the radiologist's findings, he scared the living daylight out of me. The finding was a mass, rather large—about 12 cm by 10 cm by 8 cm (5" x 4" x 3")—which they said was a mediastinum or thymic tumor, and could be cancer. The doctor wanted me to do a CT scan the very next day. However, I was so disturbed by the results and in denial that anything of that magnitude could really affect me, I put off the CT scan for three months. Besides, I had just heard somewhere that just one CT scan was comparable to six hundred chest X-rays. Wow! Wouldn't an MRI do just as well?

Apparently not, because the CT scan, although emanating ionizing radiation for about five minutes, as opposed to the thirty minutes of the magnetic radiation of the MRI, provides better details about nodules and tumors than the MRI, which only penetrates soft tissue.

Finally, at the suggestion of a retired neck surgeon who was a friend and associate of mine, I reluctantly agreed to a CT scan. My friend gently explained that he had seen so many of these masses in his day that we needed a definite diagnosis so we knew what we were up against. In my case, the CT scan was indispensable.

I agreed.

Luckily, the next day the doctor called, rather ecstatic. He said that if there was anything anyone would rather have growing in her chest, it was what I had—a cyst that was most likely benign and congenital. This cyst, however, was leaning on my trachea by nearly 45 percent and was pushing my esophagus way off and perhaps indirectly creating my hiatal hernia. After a visit to David Sugarbaker, M.D., head thoracic surgeon at Brigham and Women's Hospital, aspiration and an endoscopic removal was recommended.

And, most important for me to realize and share with you is that this single CT scan that I had eschewed privately and professionally— probably saved my life. If the cyst had continued to grow, the doctors felt it might actually damage my trachea, esophagus, and stomach because it was pushing everything out of alignment. The bottom line: some medical diagnostic tests that utilize radiation may be absolutely necessary for correct diagnosis and treatment. On the other hand, we may be unwittingly exposing ourselves to higher and higher lifetime doses of radiation by undergoing unnecessary *doctor-ordered* medical imaging, from X-rays to CAT scans to MRIs.

Reducing your exposure to nonionizing radiation such as ELFs from power lines and RF radiation from your communication devices is only part of the equation to avoid getting zapped. The truth is, according to current estimates, half of the average American's radiation exposure comes from medical imaging, a different form of radiation known as ionizing, which causes harm through heat.[1] (I say "current estimates" because those numbers could change dramatically if ELF and RF radiation were included in the studies that produced the statistics. So far, they're not because they're perceived as safe.)

So, based upon my own experience and that of many others, I want to be very responsible in weighing the benefits versus the risks of medical scans. The facts are sobering. To break it down further, an estimated 35 percent of our lifetime exposure comes from medical X-rays, and an additional 12 percent from nuclear imaging techniques which use injected radioactive materials to create images of the body. X-rays are classified as carcinogens by the World Health Organization, the Centers for Disease Control and Prevention, and the National Institute of Environmental Health Sciences because they can, in large enough doses, cause leukemia and cancers of the thyroid, breast, and lung.

It's been estimated that average Americans, who have the world's highest per-person imaging rate, are exposed to more radiation in their lifetime than workers in nuclear power plants. Think about it: an incredible 2 percent of all cancers in the United States can be attributed to medical imaging, according to a study that appeared in *The New England Journal of Medicine*. That may not sound like a lot, but it's actually one in fifty cancers![2]

Don't underestimate the risk. In a white paper published in 2007, the American College of Radiology estimated that "the current annual collective dose estimate from medical exposure in the United States has been calculated as roughly equivalent to the total worldwide collective dose generated by the nuclear catastrophe at Chernobyl."[3]

IF YOU BUY A MACHINE, THEY WILL COME

What's behind the rise in medical imaging exposures? For one thing, X-rays, MRIs, and CT scanners have been downsized, so they fit into the average physician's office. Studies have found that when doctors own MRIs or CT scanners, they're more likely to order more procedures. In one case, uncovered by a *Washington Post* investigation, within a few months

What's Your Medical Radiation Load?

Using arguably arbitrary estimates, the Nuclear Regulatory Commission has established limits to radiation exposure from medical imaging: a maximum of 50 millisieverts (mSv, a measurement of radiation absorption) for health workers in any one year and an average of 20 mSv per year. Although a patient's exposure isn't monitored, it's easy to see from the chart below just how quickly you could absorb 20 mSv in the course of diagnosing or treating just one illness or condition, particularly heart disease. The study from which this chart was developed examined insurance records of more than one million people and found that, while most were exposed to less than 3 mSv per year, about nineteen thousand received 21 mSv to 50 mSv, and almost two thousand received more than 50 mSv.[4]

Imaging Procedure	Average Radiation Dose in millisieverts
Nuclear stress test	15.6
CT angiography of chest	15.0
Inserting a coronary stent	15.0
CT of the chest	7.0
CT of the cervical spine	6.0
CT of the lumbar spine	6.0
Thyroid uptake	1.9
Mammography	0.4
Full-mouth X-ray	0.05

of purchasing a CT scanner, the number of scans ordered by doctors in a Midwest practice rose by more than 700 percent.[5] A Stanford University study looking at Medicare patients found that each new doctor's office CT scanner resulted in more than two thousand additional scans.

Convenience isn't the only factor driving that dramatic uptick. Physicians have a financial incentive to self-refer—they're reimbursed by insurance for MRIs and CT scans, money that might have otherwise gone to an outside medical imaging lab.

Medical liability is always a concern, particularly with the twin increases in medical malpractice suits and malpractice insurance costs, and that may be driving more defensive medicine procedures that protect the doctor but do nothing for you.

Boutique, or concierge, medicine—the newest trend in which patients pay a doctor a certain amount each year for old-fashioned personal attention—may also contribute to higher test rates, and therefore higher radiation levels. People who ante up thousands of dollars for this specialized care may think that taking a barrage of tests means they're getting what they pay for. In reality, they may be getting more than they bargained for.

Unnecessary tests not only expose patients to higher doses of radiation, they drive up health care costs, and lead to even more unnecessary tests if you have a false positive reading, which can also create needless worry.

Of course, the natural question that arises when you start thinking about gratuitous exposure is, how do I know when a medical test is necessary? When you go to your doctor with a symptom, your first inclination is to put the decision making into the hands of the person who actually earned a degree in medicine. And frankly, if you're concerned about a health issue, your emotional state won't allow you to switch into logical consumer mode. You're operating on fear, which is not your best companion in the doctor's office.

That's why I recommend taking someone with you whom you trust to help make the tough decisions, someone who can remain dispassionate enough to ask the right questions and take notes for you—especially if you think you are one of the growing numbers of people who may be sensitive to electromagnetic energy fields.

PARTNER WITH YOUR DOCTOR

Keep a copy of your own medical records—including your medical imaging history. I do this all the time. When you go to the doctor to discuss your results, simply ask for copies then. If you forget to do this, call the doctor's office and staff will send, fax, or e-mail your records and results to you. This is especially important if you are seeing several doctors. Instead of hard copies, you may be able to get the doctor to download copies to a flash drive, which can be plugged directly into the new specialist's computer.

One X-ray for a broken bone, CAT scan for a kidney stone, or nuclear imaging test for coronary plaque is not going to kill you. But your exposure over time does slightly increase your risk of cancer, and some factors may boost your vulnerability. For example, if you had a condition like scoliosis, a curvature of the spine, as a child and underwent many X-rays for diagnosis and then to track your treatment progress, you may be at greater risk of cancer than someone who has X-rays at an older age. Think of it this way: studies have revealed that atomic bomb survivors in Japan who were farthest from the epicenter were exposed to radiation levels comparable to the doses you'd get in just two or three CT scans, and they experienced an increased risk of cancer.[6]

Write down every X-ray or other imaging exposure you can remember since you were a child, including dental X-rays. Keep a record for everyone in the family. If you have had a significant number of exposures—more than a couple X-rays or CT scans—you will want to question your doctor carefully about the need for future tests and undergo them only when they will help you, not just when they will protect your doctor from liability or provide him or her with additional income. If you are seeing several doctors, make sure you give them the list of tests you've already had to avoid duplication and excess exposure.

Will some doctors get annoyed when you start questioning them about tests? Yes. One of my clients refused to undergo an X-ray ordered by a new physician because she had had the same test done six months previously for kidney stones and was not experiencing symptoms. The new doctor was clearly irritated and argued with her, but she stuck to

No Nukes Is Good Nukes

Living within one hundred miles of a nuclear power plant could expose you to very low levels of ionizing radiation. The safe limit? None, says the National Academy of Sciences. Even a small dose of radiation increases the risk of cancer.[7]

You may not realize it, but nuclear power plants release slightly radioactive gases and produce low-level radioactive waste. In fact, says the organization Public Citizen, nuclear plants produce an estimated two thousand metric tons of high-level and twelve million cubic feet of low-level radioactive waste in addition to fifty-four thousand metric tons of irradiated fuel annually—and it has nowhere to go. Uranium mining can seriously contaminate groundwater, and in recent years there have been leaks of a radioactive isotope called tritium into the groundwater around nuclear plants in Illinois, New York, Arizona, and Vermont. Long-term exposure can lead to cancer risk, birth defects, and genetic damage. It takes about two hundred fifty years to decay.[8]

Radiation expert Jay Gould, in his classic book *The Enemy Within,* noted that breast cancer rates quintupled in the counties that have had nuclear reactors the longest, that people who lived near nuclear facilities were more likely to have immune system disorders, and that babies born near nuclear plants were more likely to be born premature and underweight.[9] A 2009 study looking at National Cancer Institute data found an increase in leukemia deaths among children living in a county where a nuclear plant was located.[10]

If one is your neighbor (you can find them at www.insc.anl.gov/pwrmaps/map/united_states.html), regular therapeutic baths and use of zap-proof supplements may help.

her guns—then decided to find a new doctor who was more willing to treat her with respect. Later, when she was experiencing symptoms, she refused to undergo a CT involving contrast dyes (radioactive iodine or barium). The technician agreed to forgo it if she found something with the ordinary CT scan—she did, and was able to confirm the diagnosis without the use of dyes.

Image Gently for Kids

The American College of Radiology recently launched a program to encourage radiologists to follow specific procedures to give children the lowest radiation dose possible during minimally invasive interventional procedures for which imaging is used for guidance (such as biopsies).

Their advice:

- Child-size the technique: adjust the type of imaging to the size of the child.

- Use the right dose in the right place.

- Do one scan.

Share this with your child's radiologist and send him or her to www.image gently.org for more information.

At the same time, I would urge you to never refuse to have a procedure done if your doctor has legitimate reasons regarding your health. If you're anywhere but the emergency room and have the time to do so, seek out a second or third opinion.

Your first plan of action: Talk to your doctor openly and frankly about your radiation concerns. Go over the medical screening plan he or she has in mind, and follow these steps:

Ask questions. A few pointed queries can help you sort out what tests are in your best interest, the most important of which is, is this test really necessary? Ask your health care professional:

How is this going to help me? If you have a broken ankle or a kidney stone, the answer is self-evident. But some tests aren't frontline diagnostic procedures. They may not be any better at helping to identify or treat your problem than other kinds of screenings that use less or no radiation. One example is the CT scan, which exposes you to five hundred times more radiation than an X-ray. Studies have found that as many as a third of all CT scans performed in the United States were medically unnecessary.[11]

Ask your doctor if a CT scan is any better than an X-ray for your condition, or if there are any other tests that would be just as good. Your doctor should be familiar with what is called the American College of Radiology Appropriateness Criteria, which rates imaging procedures for more than one hundred fifty conditions. Scores range from 1 to 9, with 9 representing the most appropriate test for a particular disease. If your doctor tells you 1 or 2, tell him or her that you want a different kind of exam. You can view the ACR document at www.acr.org/secondarymain menucategories/quality_safety/app_criteria.aspx.

You also want your physician to tell you what he or she is going to do with the information garnered from the scans and give you the other equally or nearly as good options so you can make a more informed decision.

Is this facility accredited by the American College of Radiology? ACR accreditation ensures that the doctor and staff are well trained in operating and interpreting results from screening machines and that the devices are surveyed on a regular basis by a medical physicist to make sure they're functioning properly. The ACR also supports using the lowest doses of radiation possible to capture an image, which is more likely to be of the highest quality, eliminating the need for repeat tests.[12]

For your dentist: *Do you use high-speed X-ray films or digital imaging detectors?* Both use less radiation than conventional, slower-speed X-ray films. Don't forget to ask your dentist if X-rays are really necessary. Most adults have their mouths x-rayed once or twice a year, depending on their oral health. Children may have more screenings to check for tooth growth and hidden cavities, since they get more cavities than adults do. Like your doctor, your dentist should have a good reason to expose you or your child to X-rays, or you should refuse them.

Request a lower dose. The newest machines can be adjusted to reduce your dosage—up to about 50 percent—based on your size, though a 2001 study found that was rarely done. This is critical when scanning a child. A child's risk of developing cancer is one in five hundred from one CT scan alone. See "Image Gently for Kids."[13]

Avoid full-body scans. They offer early diagnosis, even color ultrasounds of your organs, but the so-called preventive full-body scans that claim to catch disease in its earliest stages are not only a money suck, but are dangerous. Unless you have symptoms that suggest you may have a hidden condition, there is no reason to get a scan. You're exposing yourself to a dose of radiation that is not much less than the lowest doses received by survivors of Hiroshima and Nagasaki, and you'll have a higher risk of cancer later. And about 30 to 80 percent produce some kind of anomaly that needs to be further investigated—often by more invasive or higher radiation tests—but that turns out to be nothing.[14]

Consider foregoing the EKG. Electrocardiograms, which measure the electrical activity of the heart, are used in 9 percent of routine checkups, but even with cardiac patients, the risks may outweigh the benefits. A study done by researchers at the London Chest Hospital found that an EKG done on people with stable chest pain (angina) didn't help predict who would go on to later have heart attacks or other problems.[15] The U.S. Preventive Services Task Force does not recommend using EKG or electron beam computerized tomography to screen for heart disease in low-risk adults who have no symptoms of heart disease and didn't find any evidence for or against using them in people at high risk of heart disease.[16]

Limit ultrasounds during pregnancy. There's no known risk to having one or two ultrasounds (using sound waves to create an image) during pregnancy (two is the national average). In fact, your doctor will gain valuable information about your unborn child, such as gestational age, whether the baby is growing normally, and if you're expecting more than one. But some studies in animals found that prolonged and frequent use of these devices, which expose the fetus to a 3.5 to 5.0 MHz electrical field, can cause abnormalities in fetal brain development, later behavior, and body weight.[17] Children are much more vulnerable to the effects of pollution and radiation in part because their cells are rapidly dividing and their organs and other systems are immature. While it's tempting, avoid treating ultrasound as baby's first photo op.

Protect yourself. Many years ago, I was fortunate enough to study with a remarkable woman, Dr. Hazel Parcells, an alternative medicine pioneer who foresaw the then-hidden toxic effects of this web of electromagnetic radiation we're in 24/7. She was a strong believer in long, leisurely therapeutic baths—enough hot water to draw toxins out, and cool water to remove them from the skin. Based upon her experience with the Manhattan Project and the developers of the atomic bomb in the 1940s, she recommended one therapeutic bath in particular for use after radiation exposure of any kind, including going through airport security and X-ray screenings (like mammography). If you live within a fifty-mile radius of a nuclear power plant, she recommended these baths on a bi-weekly basis. I have found this particular detox bath works wonders for relaxation and balancing if I've spent too much time on the phone or computer.

Here's what is recommended:

- Dissolve 1 pound of sea or rock salt and 1 pound of baking soda in a tub of water as hot as you can bear. (Pregnant women and those with chronic illnesses should seek advice from a health care professional before taking this bath; talk to your doctor about the appropriate water temperature for children.)

- Remain in the bath until the water is cool.

- While in the bath, sip a glass of warm water into which ½ teaspoon of rock salt and ½ teaspoon of baking soda have been dissolved. (Optional)

- Allow at least four hours to pass before showering.

Make food and supplements your radiation shield. My Zap-Proof Superfoods and Supplements were chosen specifically because they protect the body from the effects of a radiation assault. If you're undergoing medical imaging or therapy that exposes you to radiation, make sure they're part of your daily or weekly diet, particularly tart cherries, miso, whey, and supplements like melatonin, glutathione precursors and the super enzymes SOD and catalase, which are radioprotectants—proven to protect against the effects of radiation—as well as vitamin D.

You can read all about them—and sample the delicious recipes—in the next chapter.

Test for Radon

Most soils contain uranium and radon, a radioactive gas that is the by-product of uranium's natural decay. It can seep into your home from cracks in the foundation and can build up inside. It can also contaminate well water. By some estimates, it's responsible for our greatest exposure to radiation in our daily lives.[18]

You can't see or smell radon, nor does it cause any minor symptoms, but it can get into your lungs, causing damage that can lead to lung cancer, particularly in smokers. Radon is actually the second leading cause of lung cancer in the United States, and scientists are more confident about the cancer risk posed by radon than they are about other cancer causes.

Radon abatement, however, is very effective. You can reduce any radon in your home by up to 99 percent, in some cases just by fixing foundation cracks and having a vent fan installed by a qualified radon mitigator.

But first, you have to detect it. There are a number of test kits available, some at your local hardware store. There are two ways to assess your radon risk. The so-called short test involves installing a detector in your home for anywhere from two to ninety days, since radon levels vary from season to season and day to day. With the short test, you'll probably have enough of an idea of your exposure to decide if you need mitigation.

If your home measures 4 picocuries (pCi/L) or higher, you'll want to make some fixes. Even measurements lower than 4 pCi/L pose risks. If yours is high or borderline, you may want to run a second short-term test or go to a long-term test in which detection devices remain in place for more than ninety days. The U.S. Environmental Protection Agency recommends that you fix your home if the long-term test or the average of two short-term tests is 4 pCi/L or higher.

When you test your well water, make sure you test for radon. There are point-of-entry treatments that will remove the radon before it enters your home. Faucet filters are generally not as effective.[19]

See the resource guide to learn how to find a qualified radon expert in your area.

Radon Risk If You Smoke

Radon Level in picocuries	If 1,000 people who smoked were exposed to this level over a lifetime …	The risk of cancer from exposure compares to …	What to do: Stop smoking and …
20	About 260 would get lung cancer	250 times the risk of drowning	Fix your home
10	About 150 would get lung cancer	200 times the risk of dying in a home fire	Fix your home
8	About 120 would get lung cancer	30 times the risk of dying in a fall	Fix your home
4	About 62 would get lung cancer	5 times the risk of dying in a car crash	Fix your home
2	About 32 would get lung cancer	6 times the risk of dying from poison	Consider fixing between 2 and 4 pCi/L
1.3	About 20 would get lung cancer	Average indoor radon level	Reducing below 2 pCi/L is difficult
0.4	About 3 would get lung cancer	Average outdoor radon level	Reducing below 2 pCi/L is difficult

(continued)

Radon Risk If You've Never Smoked

Radon Level	If 1,000 people were exposed to this level over a lifetime....	The risk of cancer from exposure compares to ...	What to do:
20	About 36 would get lung cancer	35 times the risk of drowning	Fix your home
10	About 18 would get lung cancer	20 times the risk of dying in a house fire	Fix your home
8	About 15 would get lung cancer	4 times the risk of dying in a fall	Fix your home
4	About 7 would get lung cancer	The risk of dying in a car crash	Fix your home
2	About 4 would get lung cancer	The risk of dying from poison	Consider fixing between 2 and 4 pCi/L
1.3	About 2 would get lung cancer	Average indoor radon level	Reducing 2 pCi/L is difficult
0.4		Average outdoor radon level	Reducing below 2 pCi/L is difficult

Source: "A Citizens Guide to Radon," publication of the U.S. Environmental Protection Agency.

ZAP-PROOF
SUPERFOODS AND
SEASONINGS

C elebrated scientist Louis Pasteur, who discovered that germs cause disease, is reported to have finally admitted at the end of his life, "The microbe is nothing. The terrain is everything." In other words, a healthy internal environment is your best shield against disease. The right nutrients will build a fortress of health against the harmful consequences of constant EMF exposure. You can dramatically fortify your well-being by eating zap-proof foods and seasonings that tamp down inflammation and contain powerful antioxidants that stop oxidative stress in its tracks.

As we saw in earlier chapters, even low-level EMFs reduce major sources of antioxidant firepower in our bodies—melatonin, superoxide dismutase (SOD), and glutathione—which in turn promotes free radical DNA damage in our cells. In addition, this causes changes in calcium flow in the cells, which can suppress the immune system and keep our bodies in stress mode, triggering runaway inflammation.

Based on current scientific research, my Zap-Proof Superfoods and Seasonings arm you against EMF assault not only by replacing the antioxidants and other nutrients that electropollution robs from your body, but by turning your body into an antioxidant shield that may protect you from other dangers—heart disease, diabetes, cancer, and Alzheimer's to name a few. The foundation is foods and seasonings high in antioxidants (specifically, those shown in the most sensitive studies to be the highest in bioavailable antioxidants) and anti-inflammatory compounds, or those which have demonstrated a particular ability to protect and heal from radiation damage.

I've also chosen foods that are low or relatively low on the glycemic index, a measurement of how much a given food will raise your blood sugar. Frequent consumption of high glycemic foods has been linked to obesity and diabetes, both of which are epidemic in the United States.

I've deliberately made my dietary recommendations so simple that anyone can follow them without elaborate and time-consuming charting and tracking. Here's how it works:

1. Make the 21 Zap-Proof Superfoods and Seasonings that follow part of your weekly menu. For the most part, they're common foods, readily available in your supermarket or health food store. For the ones that aren't, I'm providing information in the resource section for mail order or online purchasing. I've included serving suggestions and serving sizes that are generally double or more what is usually recommended (a cup of broccoli, for example, instead of half a cup) for some foods that are particularly valuable to combat EMF effects.

2. Aim for 5000 ORAC points a day. ORAC is an acronym for Oxygen Radical Absorbance Capacity and is a method developed by the National Institute on Aging for measuring the ability of a particular food to fight free radicals. You can actually achieve the 5000 marker with one cup of blueberries. I've included a list of some of the top ORAC foods on page 181.

3. Make sure you get the minerals you need to keep your body running efficiently and to protect you from EMF attacks on your cell membranes and biochemical reactions. See the next chapter, "Zap-Proof Minerals and Supplements."

MY TOP 21 ZAP-PROOF SUPERFOODS AND SEASONINGS

Artichokes

One cup of cooked artichoke hearts has an antioxidant capacity that earns it the number-one spot on the USDA's vegetable list.[1] It also contains cynarin and silymarin, phytonutrients (plant chemicals) that studies show protect the liver from toxins in part by stabilizing liver cell membranes. These same chemicals are found in the herb milk thistle (artichokes are also part of the thistle family), which is used in Europe as an antidote to poisonous mushrooms. But what's important for EMF protection is that silymarin, which is ten times as potent an antioxidant as vitamin E, also increases the body's production of glutathione and SOD, the liver's premier antioxidant and key enzyme diminished by EMF exposure. One study found that silymarin boosted glutathione by 50 percent.[2] It also appears to calm inflammation and aid in cell repair—both vital if you're exposed to EMFs.

Serving suggestions: Aim for a serving (1 cup or one large artichoke) or two a week. Artichokes are available frozen and jarred year-round, but don't be daunted by the fresh version, which appears in supermarkets in season March through May.

To prepare a fresh medium to large artichoke, wash under cold running water. Trim the edible stem to about one inch (it's attached to that yummy heart, so you don't want to remove all of it). Cut off about a quarter of the artichoke top. You can use scissors to trim off the thorns from the petals, although they're also edible. You can then boil, steam, grill, roast, microwave, or braise artichokes. An Italian study found that artichokes retain high levels of antioxidants whether steamed or boiled.[3] To eat, peel off a petal, dip in melted butter with lemon, and scrape the artichoke meat off with your teeth.

You can also buy baby artichokes, which are simply a smaller variety that don't have the fuzzy center. Peel off the outer bottom leaves, trim the stems, the dark green base, and the top half inch. You can cook baby artichokes the same way you do larger ones. You can serve them with dips, in omelets, filled with seasoned breadcrumbs or cold salads, or in pasta dishes, risottos, casseroles, quiches, and stir-fries.

Asparagus

This harbinger of spring contains more glutathione than any other food. Avocado comes in second and watermelon third, but you'd have to eat at least two pounds of them to get the same amount of glutathione you'll get in five asparagus spears. Glutathione is a potent scavenger of free radicals, those rogue molecules that damage cellular DNA. It can help repair damaged DNA too, as well as binding to carcinogens and removing them from the body. Glutathione also activates other antioxidants, such as vitamin C and folic acid. As a bonus, increasing your intake of glutathione can help you detox from heavy metals, pesticides, and other noxious chemicals you encounter in the environment.

Asparagus is also a significant source of the phytochemical rutin, which strengthens capillary walls. It's also a good source of selenium and zinc, which each play a role in EMF protection. Your body needs selenium to make glutathione. In a Turkish study, rats given zinc supplements and then exposed to EMFs for five minutes a day, every other day for six months, had higher levels of the antioxidant glutathione and less evidence of free radical damage than rats who didn't have zinc supplementation.[4]

Serving suggestions: Aim for one or two servings a week. Steam, roast, or sauté fresh or frozen asparagus lightly to retain its crispness, flavor, and nutrients. Serve cold, tossed with an olive-oil or walnut vinaigrette, or warm either glistening with a little olive oil or tossed with walnut pieces and bleu or feta cheese.

Blueberries

With an ORAC score of 6552, one cup of blueberries can take care of your minimum daily recommendation of antioxidants and then some. Unfortunately, wild blueberries—the tiny ones found in your frozen food section—were not on the 2007 ORAC list, but traditionally they have scored even higher. Other berries are also good substitutes, especially blackberries (5347 per cup), raspberries (4882 per cup), and strawberries (3577 per cup).

Blueberries' secret is the level of antioxidant compounds called anthocyanins, which give them their dark color (studies show they have 38 percent more than red wine). Blueberries are also high in the antioxidant vitamin C (more than a third of the recommended daily allowance in a cup) and contain kaempferol, a phytochemical that, when it's abundant in the diet, can reduce the risk of ovarian cancer by 40 percent.

Serving suggestions: Aim for a cup or two of blueberries or other berries every day. A handful makes a delicious snack, a cup a wonderful dessert, especially when mixed with yogurt. You can bake them into your favorite whole grain muffins or bread, toss them into smoothies, or make them into refreshing cold summer soups.

Cinnamon

Just a half teaspoonful can help you lower the blood sugar–boosting effect of a high-carb food. Studies also show it can help diabetics improve their ability to respond to insulin.[5] Since there's also evidence that EMF exposure may raise blood sugar, using a little cinnamon each day may help yours remain stable, especially if your exposure is sometimes out of your control, as it often is at work. As a bonus, cinnamon is also a powerful antioxidant.

Serving Suggestions: Aim for one half teaspoon per day. Sprinkle a half teaspoon or so daily on hot cereal, toast, cooked squash, sweet potatoes, lamb, or in curries. Cinnamon is also delicious in chicken dishes made in the Moroccan style with rice, raisins, eggplant, artichoke hearts, and garbanzo beans.

Cranberries

Cranberries are high on the ORAC list, as one cup of whole cranberries has a total antioxidant capacity of 9584. That's one powerful free radical scavenger. There's good evidence that they can prevent the growth of tumors, increase good cholesterol while lowering the bad (there aren't even any drugs that can do that), kill the *H. pylori* bacteria that causes some ulcers and stomach cancer, quash the formation of dental plaque (which can lead to inflammatory gum disease), and prevent urinary tract infections. Animal studies have found that cranberries can protect brain cells from free radical damage and may prevent the kind of cognitive and even motor losses we see in the aging brain.[6]

You can actually boost the antioxidant capacity of cranberries by pairing them with apples. Apples are high in caffeic acid, which has been shown in animal studies to reduce the brain effects of EMFs from cell phones.[7]

Serving suggestions: Drink 64 ounces per day of CranWater (56 ounces of water and 8 ounces of 100 percent unsweetened cranberry juice, not cocktail!). Be aware this is a powerful diuretic and cellulite reducer, as followers of my Fat Flush diet have discovered.

Cruciferous Vegetables

High in antioxidants and vitamin C, veggies such as broccoli, cauliflower, cabbage, kale, and brussels sprouts also contain sulfur, which will boost your body's production of glutathione, the antioxidant whose production is diminished by EMFs. They're also good sources of zinc (another

EMF protectant) and selenium, and as a bonus, they're also relatively high in caffeic acid, which has been shown in animal studies to reduce the damaging effects of cell phone use.[8]

There's also strong evidence that the superstar of all phytochemicals, sulforaphane, the cancer fighter that was first isolated from broccoli, increases both the levels and activity of SOD in the body. The act of chewing a sprig of broccoli or a cauliflower floret helps activate a phytochemical called indole-3-carbinol (I3C), which helps kick-start the activity of glutathione, one of the body's cancer fighters. If you're crazy about crucifers, do eat them cooked rather than raw. When eaten raw, they can depress thyroid function in people who have hypothyroidism—too little thyroid hormone.

Serving suggestions: Aim for a minimum of three to four one-cup servings per week. Cook cruciferous vegetables until tender, as lightly as possible. Longer cooking can rob them of their vital nutrients. That smell that pervades the house when you're cooking them is those valuable sulfur compounds being released. The crunchiest of the crucifers make great crudités. With a low-calorie dip, they're a perfect snack. You can steam or sauté any of the crucifers (kale with olive oil and garlic is a wonderful side dish). Add chopped kale or cabbage to your favorite salad or soup.

7

Cumin

This peppery-citrusy spice is an important part of my Fat Flush detox plan, and it plays a vital role in my Zapped regimen as well. A powerful free radical scavenger, cumin also enhances your liver's detox antioxidant, including glutathione. In one study, it increased the activity of the glutathione enzyme by 78 percent![9] Other research has found that the essential oils of cumin, when exposed to microwave and gamma radiation (a form of ionizing radiation), actually have more antioxidant power, which suggests it could be one of your body's chief defenders against lower levels of nonionizing radiation.[10]

Serving suggestions: Aim for three servings of about $1/2$ teaspoon per week. Cumin's perfect partner is beans, specifically red beans, another

Zapped superfood. Cumin is used along with chili powder in most chili recipes. It's truly an international spice, found in Mexican, Indian, Greek, and Middle Eastern dishes. Use it in curry (including vegetable curry, so you can pair it with the healing cruciferous vegetables) or as part of a rub or marinade (garlic, lemon, and olive oil) for grilled or broiled grass-fed beef steak or chicken.

Garlic

An anti-inflammatory food, there's also some evidence that garlic can help control blood sugar, which may rise when you're exposed to EMFs. It's also high in sulfur-containing compounds, which play a role in the production of glutathione. Interestingly, these are the reason garlic is called the stinking rose—they're responsible for its pungent odor.

You've probably heard that garlic can reduce the risk of cardiovascular disease. It works in many ways to protect your heart, including inhibiting calcification (the layering of calcium) in coronary arteries, a precursor to the hardened plaque associated with poor blood flow and clotting which can lead to heart attack and stroke. Studies have also found that garlic can reduce free radicals in the bloodstream, which is probably what contributes to what other researchers have found: garlic inhibits plaque formation by up to 40 percent, likely because these hardened chunks of cholesterol and other bloodstream debris are created when cholesterol is oxidized by free radicals. A free radical scavenger like garlic can nip the entire plaque process in the bud. I usually recommend ¼ teaspoon to ½ teaspoon of aged garlic extract per day, which is equal to one to two capsules at 300 mg each.

Garlic also contains antioxidant vitamins C and E as well as selenium, an important cofactor mineral in the production of glutathione. And it works in an interesting fashion against one particular carcinogenic pathway: the cancer-causing chemicals produced when you grill meat or cook them at high temperatures. One of those carcinogens, called PhIP, may be one reason for the high rate of breast cancer among women who eat large quantities of meat. One of garlic's organic sulfur compounds prevents PhIP from becoming carcinogenic.[11] That same compound also triggers the genes that produce SOD and glutathione, which may help protect you from those cancer-causing chemicals—and EMFs.

Serving suggestions: Aim for one serving (one-half to one clove) a day. Chopping or crushing garlic causes the compound alliin to transform into allicin, to which much of garlic's health benefits have been attributed. Wait several minutes before eating or cooking garlic for that process to take place. Cook lightly—after 10 minutes of heat, garlic will lose its phytonutrient power. You can use garlic as an ingredient in salad dressings and marinades (especially for grilled meats), as flavoring for vegetables, or mixed with garbanzo beans, tahini, olive oil, and lemon to make hummus. For those who can't handle the taste of garlic (or the breath afterwards), garlic extract is a good alternative.

Grass-Fed Beef

If you eat beef, grass-fed is a must. It is an excellent source of glutathione, zinc, and selenium—all nutrients reduced by EMF exposure. And there are some great bonuses. Compared to grain-fed beef, which contains 40 percent saturated fat, cattle that graze on pasture grass have only 10 percent of the heart-threatening fat. Grass-fed beef is also higher in nutrients, including beta carotene and vitamin E, and has more heart-healthy, anti-inflammatory omega-3 fatty acids and conjugated linoleic acid (CLA), a critical fatty acid that studies suggest may help lower your risk of cancer, heart disease, diabetes, and love handles (in some research, CLA helps reduce body fat).[12]

Buy beef products marked with the logo of the American Grass-fed Association (AFA), which certifies the animals have been raised on nothing but mother's milk and forage—not corn or other grain which is often used to promote quick weight gain. The AGA seal guarantees that producers did not keep cattle confined nor use antibiotics or hormones. (On a personal note, I order all of my beef from www.ranch foodsdirect.com, which provides grass-fed beef and uses a proprietary method known as "rinse and chill" to clean the beef. See resources for more information.)

Serving suggestions: Aim for at least two 3- to 4-ounce servings per week. Grass-fed beef comes in all cuts and can be used in any of your favorite recipes. You will find the Ranch Foods Direct beef unusually tender, unlike other grass-fed beef, but it is still best to avoid overcooking to preserve taste, texture, and glutathione in the meat, which are all diminished by cooking.

Mushrooms

Mushrooms, particularly Asian varieties such as shiitake, maitake, crimini, oyster, and king oyster mushrooms, contain high concentrations of a powerful antioxidant that could help protect your cellular DNA from free radical damage and can help slow the development of chronic degenerative diseases associated with aging. In fact, Asian mushrooms contain twenty times more of the antioxidant L-ergothioneine than do wheat germ and chicken livers, the other most abundant sources. But even white button mushrooms have fifteen times more L-ergothioneine than those two foods.[13]

Mushrooms also contain lentinan, an immune-system booster, which reduced the development and size of tumors in lab animals injected with human colon cancer cells. Mushrooms are also an excellent source of selenium and copper, and a good source of zinc. Selenium contributes to the body's production of glutathione, copper is an important cofactor in SOD production, and zinc has been found to protect against free radical damage in animals exposed to 900 MHz radio waves from a cell phone.[14]

Zinc is also a vital immune-system booster and lowers blood sugar, which are both affected by EMFs.

Serving suggestions: Aim for two to three 1-cup servings a week. Don't wash mushrooms—their skins are so porous, they'll drink in the fluid. Clean by wiping with a damp cloth, and then sauté with garlic to serve with meat or over vegetables such as asparagus or broccoli, or add to pasta sauce and omelets. Make your own veggie burgers by tossing a variety of sautéed mushrooms with some beans (black or red) and sautéed onions in the food processor for a few pulses (you don't want to puree them). Add your favorite seasoning and enough whole wheat bread crumbs, cooked bulgur, or brown rice so you can form patties. Grill or pan fry.

Olive Oil

At first researchers were baffled by the conundrum (also called the French Paradox): how could Mediterranean people who ate more fat than Americans be so much healthier? As it turned out, it's not about how much but what *kind* of fat you eat. Studies suggest that relying only on olive oil can cut your risk of heart disease almost in half and your chances of dying prematurely by a full 50 percent.[15] But what's important for our purposes is that olive oil is a potent antioxidant that boosts levels (to higher than normal!) of the two forms of glutathione—reductase and peroxidase—that can protect you against free radical cellular damage.

Choose extra virgin olive oil (the least processed) for its oleic acid, which is anti-inflammatory and may help reduce arthritis and asthma symptoms and preserve bone density. In one study, in fact, four tablespoons of olive oil produced a pain-relieving effect that was equivalent to about 10 percent of the typical adult dose of ibuprofen.[16] It also helps control blood sugar, which can be affected by exposure to RF.

Serving suggestions: Aim for at least 1 tablespoon a day. Except in baking, whenever a recipe calls for oil, use extra virgin olive oil. You can toss whole wheat or gluten-free pasta or rice with olive oil, parmesan cheese, and garlic; use it in place of butter or margarine on bread and vegetables, and when cooking meat (it reduces the production of carcinogenic compounds in cooked meat, particularly when you combine it with rose-mary). Make sure you keep your olive oil in the dark. Light quickly reduces its potency. In one Italian study, after just two months of exposure to a su-permarket light, olive oil lost 30 percent of its vitamin E and carotenoids, and had high levels of free radicals (which, when it comes to oil, means it's becoming rancid).[17] Tinted glass bottles are best for storage.

Pomegranate Juice

Red wine takes a nutritional backseat to this relatively new and ultra popular drink with an ORAC value of 2341 for about 2 ounces. A study at the University of California at Davis found that the antioxidant activity of commercial pomegranate juice is three times higher than that of red wine and green tea. In fact, the fruit juice neutralized 54 percent more free radicals than the much-heralded wine.[18] People who drank about 2 ounces a day had an average increase by 9 percent in antioxidant ac-tivity, according to another study.[19] In other research, pomegranate juice was better able to prevent oxidation of LDL (bad) cholesterol. Oxidation is what triggers cholesterol to clump and stick to artery walls. The juice also increased blood flow to the heart in forty-five people with heart dis-ease who drank eight ounces a day for three months.[20]

Serving suggestions: Aim for two or three 8-ounce servings of juice per week (which I would dilute half and half with water to reduce the concentration). Studies have focused primarily on the juice rather than the fruit, so you can use pomegranate juice as a foundation for healthy smoothies, or in cooking as a marinade. There is even a pomegranate wine from Rimon Winery in Israel.

Prunes (Dried Plums)

With an ORAC rating of 6552 for 3.5 ounces, prunes (now called dried plums for public relations reasons) contain unique killer antioxidants. They're especially effective against a very dangerous free radical called superoxide anion radical—the main target of SOD. They prevent free radicals from causing damage to fats, which are essential to cell membrane and brain cells; they also protect against peroxidation, the harmful effects of oxygen on cholesterol, which can trigger the cascade of events leading to plaque formation and atherosclerosis, a risk factor for heart attack and stroke. Prunes are also a significant source of vitamin A (as beta carotene), another antioxidant that protects the integrity of cell membranes and is also an anti-inflammatory. Because they contain soluble fiber, prunes can also help you keep your blood sugar—which is affected by low-level radiation—on an even keel.

Serving suggestions: Aim for two to three servings (a serving is two medium-size prunes) a week. Stuffed with an almond or walnut, prunes make a delicious snack that's as satisfying as candy, especially for children. The dried fruit will soak up any marinade, so add it to Middle Eastern chicken and beef dishes or desserts. Of course, pureed prunes make a wonderful nutrient-rich sugar substitute in baking.

Red Beans

Red beans, including small red, red kidney, and pinto beans, came up tops (with a rating of 8459) in a 2007 study of antioxidant absorption of foods done at the USDA's Arkansas Children's Nutrition Center in Little Rock. That's important because, while other foods may have higher antioxidant capacity, they're not as readily absorbed as beans. The darker the better: their color reflects their content of phenol and anthocyanin antioxidants. A great source of protein, especially on healthy vegetarian diets, beans are high in fiber (one cup gives you about 45 percent of your daily fiber needs) and low on the glycemic index. Their soluble fiber actually helps stabilize blood sugar and, because beans are so high in thiamin, a cofactor in the production of the memory-linked brain chemical acetylcholine, they can also protect your brain cells.

Serving suggestions: Aim for three cups a week. Beans need to be cooked for a long time, which does leach out some of their nutrients. But that also means that dried and canned beans have about the same

Worth the Trouble

These are foods that aren't necessarily going to be in your local supermarket, but they're usually available in your health food store or online. They are so beneficial to your overall health that I think it's worth the time and trouble it takes to find them.

Açaí

This little berry from the Amazon rain forest has taken the health food industry by storm. Unlike many other foods du jour, açaí is the real deal. It's not a miracle food, but it is a big free radical fighter in a very small package. Its ORAC ranking depends on how you consume it—freeze-dried açaí has been measured at 161,400, while the mixed juice blend has a respectable 5500. The berry pulp is available frozen. You just can't get it fresh: the berries are so fragile, they can't survive shipping. But that's great, because you can use frozen açaí berry pulp in smoothies and desserts or mix it with yogurt. Like blueberries and other dark-colored berries, it contains powerful antioxidants called anthocyanins.

Mangosteen

A delicious tropical fruit, mangosteen—like açaí—suffers from overhype on the Internet, which obscures its real value as an antioxidant and potential cancer fighter. Japanese scientists found that it acted as an anti-inflammatory, much like COX–2 inhibitors like Celebrex, particularly in the structural cells of the brain, making it potentially a preventive for conditions such as Alzheimer's disease which may be caused by exposure to EMFs.[21]

Noni

The juice of this Polynesian fruit (from an evergreen) has been shown in laboratory testing to be a potent antioxidant and an immune stimulator as well as a protectant against cancer. Studies at the University of Hawaii's Cancer Research Center found that noni, given in capsules containing freeze-dried extract, also reduces pain.[22]

nutritional value, so you don't need to go through all that soaking and cooking yourself. Canned beans are fine—just rinse thoroughly to remove all the sodium. You can use red beans in ethnic dishes such as chili, enchiladas, and burritos; as a base for veggie burgers; or cold, tossed with other beans, garlic, and other zap-proof seasonings like rosemary in three-bean salad for multiple EMF protection.

Rosemary

In studies looking at the damage caused by gamma radiation, Indian researchers found that rosemary protects cellular DNA from damage in several ways, including acting as an antioxidant.[23] Rosemary-treated groups of mice exposed to radiation had an increase in the number of disease-fighting white blood cells called leukocytes. They also showed a significant decrease in oxidative degradation of blood fats—called peroxidation, a major marker of cardiovascular disease—and an increased level of glutathione.[24] That's important protection because radiation causes an increase in free radical damage to lipids and drops glutathione levels. Rosemary also gives vitamin E a boost so it can continue scavenging free radicals over and over again, something rosemary can do without any help. It's such an effective antioxidant, the food industry uses rosemary components as a food preservative. As a bonus, studies also show that it can reduce heterocyclic amines, carcinogenic compounds that form when meat is cooked at high temperatures.

Serving suggestions: Aim for two to three 1-tablespoon servings a week. Rosemary gives flavor and oomph to meats, egg dishes, and salad dressings. It's also a perfect seasoning for pasta sauces.

Sea Vegetables (Seaweeds)

McGill University researchers found that alginic acid, which is found in brown algae like kelp and alaria, reduced the amount of strontium 90—

one of the most common radioactive materials in the environment—absorbed through the intestinal wall.[25] Because it is so widely dispersed these days, in part from residual fallout from worldwide nuclear testing, most of us are exposed to strontium 90 in food or water or in the dust we inhale. Other sea vegetables (like nori, hijiki, arame, kombu, sea palms, and wakame) are also rich in iodine, which can keep your thyroid—a target of radiation, including EMFs—healthy. Sold in natural food stores throughout the country, these sea veggies are packaged in a dried form with serving suggestions on the package.

Serving suggestions: Straight from the package or slightly toasted in the oven, nori can be tossed and crumbled over a variety of dishes. It is delicious over vegetables, whole grain pasta, or fish. Hijiki can be added to soups or salads and tastes great sautéed with carrots and fresh ginger. Wakame adds to soups, on top of fish, or in dressings. It also makes a great marinated salad with cucumber and apple cider vinegar. Kombu can be toasted in the oven for snacks, added to beans (aids digestion), and cut into strips to be added to soups.

Tart Cherries

Montmorency cherries, the most common tart cherries produced in the United States, contain significant quantities of melatonin, the antioxidant hormone produced by the pineal gland that is targeted by EMFs. In fact, they contain even more than is normally found in the blood. That was the surprising discovery made recently by the University of Texas Health Science Center's Dr. Russel Reiter, who has been studying melatonin for more than thirty years.[26] Studies have confirmed that melatonin is radioprotectant, a substance that either lessens or prevents the health effects of radiation, largely by preventing damage caused by free radicals.

Melatonin plays a role in the production of the body's own potent free radical scavengers, glutathione and SOD. It also rules our circadian rhythms, which supply us with chemicals that allow us to sleep and encourage us to wake up. Sleep experts confirm that most Americans,

High ORAC Foods

Apples		**Cabbage, red**	2252
Golden Delicious	2679	**Cranberries**	9584
Red Delicious	4275	**Currants, red**	3387
Granny Smith	3898	**Figs**	3383
Artichokes	6552	**Juice**	
Asparagus	2150	Concord Grape	2377
Basil	4805	Pomegranate	2341
Beans		**Nuts**	
Pinto Beans	7779	Hazelnuts/filberts	9645
Red Beans	8459	Pecans	17,940
Berries		English walnuts	13,541
Blackberries	5347	Peanuts	3166
Blueberries	6552	**Plums dried**	6552
Broccoli	2386	**Raspberries**	4882

even those who don't travel, are effectively suffering from jet lag, a sleep deficit so severe it may be contributing to a variety of illnesses, including frequent colds and viral infections, obesity, diabetes, and heart disease. Cherries are also anti-inflammatory and relatively low on the glycemic index.

Serving suggestions: Aim for two to three servings a week of 8 ounces of cherry juice or 3.5 ounces of dried cherries, more if you're not taking melatonin supplements. I would dilute the juice half and half with water. Tart cherries are often used in pies, but that's not the healthiest delivery system for a food that can protect you from the effects of EMFs. They're also sold dried, which makes a handy, right-from-the-bag snack that's actually higher in melatonin than fresh cherries. Consider dried cherries as an addition to healthy morning muffins, gluten-free cereal, oatmeal, gluten-free pancakes, salads, as well as rice, rice pasta, buckwheat, amaranth, and quinoa. Mix a bit of tart cherry juice (regular or concentrated) with water, particularly when you're exercising. One study found that it may reduce joint inflammation caused by physical activity.[27] If you don't like or don't want to eat cherries, there are several other foods that contain small amounts of melatonin, including bananas, onions, corn, oats, and rice.

Turmeric

This spice, used to color and flavor the mustard you slather on your ballpark hot dog as well as Indian curry, has been linked to lower risk of leukemia—because it inhibits radiation-induced chromosome damage![28] Turmeric, whose chief phytochemical is called curcumin, also protects against damage caused by other environmental pollutants as well as the carcinogens in cooked meat. In test tube studies, it halted the proliferation of leukemia cells. It also has been shown to boost the detoxing ability of the liver enzymes, including glutathione, as well as blocking free radical damage on its own.[29] Turmeric has an ORAC value of 2117 per 3.5 ounces.

Neurodegenerative diseases have been linked to EMF exposure, and turmeric, which crosses the blood-brain barrier, shows powerful promise in fighting them. In studies it slowed the progression of an Alzheimer's-like disease in animals,[30] boosted the production of other antioxidants, and helped prevent the formation of so-called amyloid plaques that are the hallmark of Alzheimer's, while it also cleared existing plaques from the brain.[31]

As an anti-inflammatory, turmeric has outperformed prescription and over-the-counter drugs. Interestingly, research has also found that turmeric makes cruciferous vegetables even more potent anticarcinogens. Alone, neither curcumin nor the phytochemical phenethyl isothiocyanate—from cabbage, broccoli, and the like—exerted any effect on prostate tumor cells; together, they stopped them dead.[32]

Serving suggestions: Aim for at least 1 tablespoon per day. Use real turmeric rather than curry powder, which doesn't contain enough of the spice to be physiologically active. Add a teaspoon or so of turmeric to your favorite bean dishes, salad dressings, curries, and absolutely on cruciferous vegetables—it's especially good on cauliflower, which you already know if you're a fan of Indian food. Combine it with cinnamon, cumin, and other spices to coat chicken or fish. Mix your own cruciferous veggie dip using plain Greek yogurt spiced with turmeric, lemon juice, and a little dried minced onion and garlic powder, and even a little horse-radish sauce for heat. The turmeric will give it a lovely yellow color. Add some chutney for a delicious sauce for meat, chicken, or fish.

Wild Alaskan Salmon

Along with other coldwater fatty fish, salmon is rich in anti-inflammatory omega-3 fatty acids. Omega-3s boost liver function to burn fat and detoxify the body, reset healthy levels of the brain's neurotransmitters that regulate appetite, and strengthen cell membranes while optimizing cell function so cells keep flushing out waste and taking in nutrients. They also exert a protective, even healing effect on the brain. Studies have found that children with learning and behavior problems show improve-ment after fish oil supplementation.[33] Animal studies have suggested that omega-3s may also help prevent neurological disorders such as Parkin-son's and Alzheimer's diseases. I recommend wild Alaskan salmon spe-cifically because it contains fewer pollutants than other kinds of salmon. Canned salmon is usually wild salmon.

Other safe and eco-friendly fish alternatives: Pacific sardines, light tuna, rainbow trout; Northern, Japanese, or European anchovies; and Pacific halibut. If you don't like fish, choose other high omega-3 foods, such as walnuts, chia seeds, Perilla oil, or flax seeds. (An omega-3 power-house, flax seeds are rich in its precursor, alpha-linolenic acid. Flax seed oil contains a more concentrated source.)

Serving suggestions: Aim for one to two fatty fish meals a week, with salmon as at least one of them. Use canned salmon in salads (mix with a little light mayo and dill for a fabulous lunch) or grill, poach, broil, or bake salmon fillets or steaks. Consider supplementing with 2 to 4 grams of fish oil capsules daily.

Walnuts

Make walnuts your protein snack of choice. They're the only nut that supplies significant amounts of omega-3 fatty acids, a natural anti-inflammatory, through alpha-linolenic acid, which becomes an omega-3 in the body. They are also high in glutathione—they're actually the only nut that contains this vital antioxidant.

A USDA study now under way is looking at the potential of walnuts to protect nerve cells in the brain that degenerate with aging and in disorders such as Alzheimer's and Parkinson's diseases, which have been linked to EMF exposure. Walnuts have a total ORAC rating of 13,541 per 3.5-ounce serving (pecans are higher, and are a good substitute, though you'll be missing out on the omega-3s), which may be responsible for their heart-protective properties.

Serving suggestions: Aim for a handful of walnuts a day. Toss them into your favorite salad, tuck a walnut half into a prune for a terrific snack, add them to stir-fries, or use them to make pesto (with basil, parmesan, and olive oil, and maybe a little yogurt to make a creamy sauce, or in my Walnut Rosemary Pesto on page 197).

Yogurt

One theory explaining the symptoms of electrosensitivity and radio wave sickness is that they are caused by the loss of calcium ions due to EMF exposure. Since these ions hold the membranes together, the cell is more vulnerable to damage. Dr. Andrew Goldsworthy was the first to notice the similarity between electrosensitivity symptoms and those of hypocalcemia, a deficit of calcium in the blood. He theorized that if someone already had low blood calcium, EMF exposure, which leaches calcium from the cells, might "push them over the edge."[34] The treatment: calcium. However, calcium must be in balance with magnesium for proper absorption. The ideal ratio should be at least 1:1 or even 2:1 in favor of magnesium.

When it comes to calcium, food is your best source. I like yogurt because it's easily digestible, less likely to cause allergic reactions, can be safely eaten by those who are lactose intolerant, and contains probiotics—beneficial bacteria—that are vital to a healthy digestive and immune systems. Yogurt is also an unexpected source of iodine, which can help promote better thyroid function. (Animal studies have found that radiation can reduce levels of the thyroid-stimulating hormone.)

Serving suggestions: Mix your high-ORAC fruits with organic yogurt as a breakfast food, part of a smoothie, or a delicious dessert. Yogurt is a fabulous base for dips (for those cruciferous vegetables you're eating and even fresh strawberries), salad dressings, sauces, and in place of milk in your favorite recipes. Mix with frozen berries (like açaí, available in health food stores) for a refreshing treat.

RECIPES

I tweaked a few of my favorite recipes to show you how effortlessly you can integrate my Zap-Proof Superfoods and Seasonings into your everyday diet. These recipes are brimming with flavor and bursting with health benefits that will deliciously fortify your system.

Zippy (not Zappy) Berry Smoothie

MAKES 1 SERVING

This is the best and easiest breakfast ever. The fruit and juice give you an early morning antioxidant boost, the yogurt provides calcium, the flax extra protection from cancer, and the whey powder—along with protecting you from harmful EMFs—fills you up so you're not hungry all morning. Add your favorite berries or fruit, and juice.

½ cup fresh or frozen berries
½ cup unsweetened cranberry juice, tart cherry juice, or pomegranate juice
4 ounces plain yogurt of your choice (Greek yogurt is ultra creamy!)
1 tablespoon flaxseed oil or ground flaxseed
1 scoop Fat Flush Vanilla Whey Protein Powder
1 cup ice

Place ingredients in blender. Blend until creamy. Add a little more juice or water if the smoothie is too thick.

Per Serving, About: Calories: 340 Protein: 26 g Carbohydrates: 20 g Dietary Fiber: 3g Sugars: 13 g Total Fat: 15 g Saturated Fat: 3 g Cholesterol: 30 mg Sodium: 140 mg Potassium: 340 mg

Coffee Smoothie

MAKES 1 SERVING

1 cup organic coffee
1 teaspoon Flora Key natural probiotic sweetener
6 ounces vanilla yogurt
½ small banana, cubed
¼ teaspoon cinnamon
1 cup ice

Add natural probiotic sweetener to coffee after it cools off a bit. Then blend all ingredients until smooth.

Tip: Coffee is high in caffeic acid, shown to protect against harmful cell phone radiation. Consider adding a tablespoon of instant coffee, mixed with a tablespoon of hot water, in your favorite chili or stew recipes. You won't taste the coffee at all—it will just give your meal a rich flavor.

Alternatives: Add a pinch of cocoa powder for café mocha; freeze the banana and cut into chunks; or add a tablespoon or two of chocolate whey protein powder.

Per Serving, About: Calories: 200 Protein: 7 g Carbohydrates: 30 g Dietary Fiber: 3 g Sugars: 23 g Total Fat: 0 g Saturated Fat: 0 g Cholesterol: 0 mg Sodium: 90 mg Potassium: 500 mg

ENTREES

Old-Time Western Chili

MAKES ABOUT 8 SERVINGS

1 tablespoon extra virgin olive oil
2 pounds grass-fed stew beef, cut into cubes
1 medium yellow onion, diced
1 15-ounce can green chilies
2 cloves garlic, mashed
1 red bell pepper, diced
1 medium jalapeño, diced
3 15-ounce cans red kidney beans, rinsed
1 10.5-ounce can unsalted tomato puree
2 tablespoons chili powder
2½ teaspoons cumin
3 teaspoons dried oregano
Salt

Place the oil in an oven-proof skillet on the stove top and brown meat. Add onions, green chilies, garlic, red pepper, and jalapeño and sauté until the onions are translucent.

Add tomato puree and chili powder, cumin, and oregano. Bring to a simmer. Cover the pot and place it in a 325°F oven and cook for 1½ to 2 hours until meat is tender. Stir occasionally and add water if it gets too thick. Add beans and return to oven for 15 minutes. Salt to taste.

Tip: For a rich-tasting antioxidant boost, add a tablespoon or two of natural, unsweetened cocoa powder. Cocoa contains more anti-oxidants than any other chocolate product. Don't worry, your chili won't taste like hot chocolate. The cocoa just adds a more complex, hearty flavor. It's similar to traditional Mexican mole poblano sauce, which combines a little chocolate with a variety of chilies and other ingredients.

Per Serving, About: Calories: 415 Protein: 30 g Carbohydrates: 36 g Dietary Fiber: 12 g Sugars: 4 g Total Fat: 18 g Saturated Fat: 4 g Cholesterol: 68 mg Sodium: 487 mg Potassium: 715 mg

Grass-Fed Beef Steak with Garlic, Wine, Rosemary, and Exotic Mushrooms

MAKES ABOUT 8 SERVINGS

2 cloves garlic

1¼ teaspoons dried rosemary leaves or 2 tablespoons fresh rosemary

½ cup dry red wine

2 tablespoons extra virgin olive oil

2 pounds sirloin steak, about ½ inch thick

4 cups exotic mushrooms (1 cup each portobello, crimini, oyster, and enoki)

Salt

Place the garlic, rosemary, and half the red wine in food processor and pulse until mixed thoroughly, adding half the olive oil to make a paste. Coat both sides of the steak and marinate for a minimum of 2 hours, preferably overnight. Grill or broil steak at medium-high for about 3 minutes each side for medium rare. Sauté mushrooms in remaining red wine and olive oil and serve with steak sliced on the grain. If there are any leftovers, toss them in a salad the next day.

Tip: Grass-fed beef has less fat than grain-fed meats, so you may need to use a little extra virgin olive oil for browning. This also means that it's easy to overcook grass-fed beef. If you like your steaks well-done, cook them at lower temperatures in a sauce, marinade, or rub. It's also a good idea to sear the meat first at high temperatures to seal in juices.

Per Serving, About: Calories: 190 Protein: 25 g Carbohydrates: 3 g Dietary Fiber: 1 g Sugars: 1 g
Total Fat: 7 g Saturated Fat: 0 g Cholesterol: 60 mg Sodium: 65 mg Potassium: 23 mg

Hearty Beef Stew

MAKES 4 SERVINGS

1 pound grass-fed beef stew meat (lean bottom or eye round for quicker
   cooking)
Gluten-free whole grain flour, such as brown rice flour, for dredging
1 tablespoon extra virgin olive oil
1 large onion, coarsely chopped
2 large red, yellow, or orange bell peppers, coarsely chopped
3 celery stalks, finely chopped
4 garlic cloves, minced
1 10.5-ounce can tomato puree
⅓ cup dry red wine
1½ teaspoons fresh oregano
1 ½ teaspoons fresh basil
1 bay leaf
1 large or 2 medium carrots, sliced
2 small potatoes, peeled and chopped into 1-inch cubes
2 tablespoons chopped parsley
1 cup mushrooms, sliced
Salt

Cut stew meat into one-inch cubes and dredge in flour. Heat oil in
large Dutch oven and sauté meat, onion, peppers, celery, and garlic.
Stir in tomato puree, wine, and remaining seasonings and enough
water to cover. Simmer for about an hour. You may need to add a little liquid during the cooking process to prevent the stew from getting too thick. Add carrots, potatoes, parsley, and mushrooms and
simmer for another 30 minutes till meat is tender. Don't cook for
much longer—grass-fed beef can get dry and tough if overcooked.
Salt to taste and serve.

———————

Per Serving, About: Calories: 500 Protein: 27 g Carbohydrates: 35 g Dietary Fiber: 8 g Sugars:
10 g Total Fat: 20 g Saturated Fat: 4 g Cholesterol: 68 mg Sodium: 400 mg Potassium: 1,300 mg

Grilled Wild Salmon with Mango-Pomegranate Salsa

MAKES 4 SERVINGS

SALSA INGREDIENTS

1 mango, ripe, diced

¼ cup chopped red or yellow bell pepper

¼ cup diced red onion

½ cup pomegranate arils (seeds)

1 teaspoon extra virgin olive oil

1 tablespoon diced jalapeño or serrano pepper (or to taste)

2 tablespoons chopped fresh cilantro

1 teaspoon ground cumin

½ avocado, diced

1 lime, juiced

SALMON

4 wild Alaskan salmon fillets (5 ounces each)

Extra virgin olive oil

1 lemon or lime

To make salsa: Place the mango, pepper, onion, pomegranate seeds, and olive oil in a bowl.

Add a small amount of jalapeño or other hot pepper and taste. Add more until it reaches desired hotness. Add cilantro, cumin, avocado, and lime juice. Toss.

To make salmon: Brush salmon fillets with olive oil. Prepare grill or broiler.

Season fish with juice of lemon or lime to personal taste. Grill or broil till fish is no longer translucent—about 8–10 minutes. Transfer fish to serving plate and top with salsa. Serve immediately.

———

Per Serving, About: Calories: 330 Protein: 25 g Carbohydrates: 18 g Dietary Fiber: 4 g Sugars: 12 g Total Fat: 15 g Saturated Fat: 3 g Cholesterol: 60 mg Sodium: 70 mg Potassium: 700 mg

VEGETABLES

Roasted Asparagus

MAKES 4 SERVINGS

Roasting vegetables brings out their inherent sweetness. Even non-veggie lovers will devour this dish.

1 pound asparagus spears (thicker is better for roasting)
1–2 tablespoons extra virgin olive oil
2 garlic cloves, minced
Lemon juice
Salt

Preheat oven to 400°F. Clean asparagus and break off the tough ends.

Mix olive oil, garlic, and a squeeze of lemon juice. Toss asparagus in olive oil mixture. You can also place the olive oil mixture and asparagus in a plastic bag and rub the bag so the spears are thoroughly coated. Place spears in a single layer in a shallow roasting pan or glass dish. Season with salt. Place in oven and cook for anywhere from 10 to 20 minutes, depending on your preference. The longer you cook the asparagus, the crispier the ends will be.

Per Serving, About: Calories: 55 Protein: 4 g Carbohydrates: 4 g Dietary Fiber: 4 g Sugars: 4 g Total Fat: 3 g Saturated Fat: 0 g Cholesterol: 0 mg Sodium: 4 mg Potassium: 7 mg

Brussels Sprouts

MAKES 4 SERVINGS

1 pound brussels sprouts
2 tablespoons extra virgin olive oil
3 cloves garlic, minced
1 tablespoon balsamic vinegar
Salt

Preheat oven to 450°F. Trim the ends off the brussels sprouts and re-move any discolored leaves. Cut sprouts into halves or quarters. Put them in a bowl and toss with olive oil, garlic, and balsamic vinegar. Arrange sprouts in a single layer in a shallow roasting pan or glass dish and roast for 20–25 minutes, turning occasionally, until they're slightly crisp. Season with salt.

Serving suggestion: Toast some chopped walnuts and toss with roasted brussels sprouts.

Tip: Sweeten up bitter vegetables with balsamic vinegar, which is made from the juice of the Trebbiano grape. In Italy, it's actually used in desserts! Try it on some fresh strawberries, pears, or tomatoes.

Per Serving, About: Calories: 111 Protein: 4 g Carbohydrates: 13 g Dietary Fiber: 4 g Sugars: 4 g Total Fat: 6 g Saturated Fat: 0 g Cholesterol: 0 mg Sodium: 28 mg Potassium: 12 mg

Cabbage-Cranberry Sauté

MAKES 4–6 SERVINGS

2 tablespoons extra virgin olive oil
1 red onion, thinly sliced
1½ pounds red cabbage, thinly sliced
¼ cup unsweetened pomegranate, tart cherry, or cranberry juice
2 tablespoons dried and unsweetened cranberries
½ cup apple cider vinegar
½ teaspoon turmeric
Salt

Heat oil in a large skillet and sauté onion until translucent. Add cabbage and cook until soft. Pour in juice, and then stir in cranberries, vinegar, and turmeric. Cook until sauce reduces and simmer for 20 minutes. Salt to taste.

———————

Per Serving, About: Calories: 100 Protein: 1 g Carbohydrates: 13 g Dietary Fiber: 3 g Sugars: 8 g Total Fat: 4 g Saturated Fat: 0 g Cholesterol: 0 mg Sodium: 34 mg Potassium: 71 mg

Artichoke and Roasted Red Pepper Salad with Rosemary Walnut Pesto

MAKES 4 SERVINGS

This is a quick and easy cold side dish (though you can heat it up) and makes a nice light lunch or dinner.

    1 14-ounce can artichoke hearts, drained (frozen artichokes may also be used)
    1 12-ounce jar roasted red peppers
    1 red onion, chopped
    Salt
    Fresh parsley, for garnish

Combine all ingredients in a bowl. Mix in Rosemary Walnut Pesto (see following recipe). Salt to taste and garnish with fresh parsley. Serve.

Rosemary Walnut Pesto

1½ cups walnuts
¼ cup fresh rosemary
1 clove garlic
⅓ cup extra virgin olive oil
¾ cup grated Parmesan or other hard cheese

Place whole walnuts in a shallow roasting pan coated with cooking spray and place in a 350°F oven. Bake until golden, 5–10 minutes. Place walnuts, rosemary, and garlic into food processor and blend, gradually adding the olive oil in a stream. When the pesto is thick and smooth, remove from processor and stir in cheese.

Per Serving, About: Calories: 526 Protein: 20 g Carbohydrates: 15 g Dietary Fiber: 7 g Sugars: 4 g Total Fat: 45 g Saturated Fat: 5 g Cholesterol: 16 mg Sodium: 700 mg Potassium: 80 mg

Piquant Veggies

MAKES 4–6 SERVINGS

½ head broccoli broken into bite-size florets
½ head cauliflower broken into bite-size florets
1 6-ounce jar marinated artichoke hearts (do not drain)
1 lemon
Salt

Steam broccoli and cauliflower until tender crisp. Toss with marinated artichoke hearts. Squeeze lemon on top. Heat in skillet until hot.

Per Serving, About: Calories: 77 Protein: 2 g Carbohydrates: 8 g Dietary Fiber: 2 g Sugars: 2 g Total Fat: 5 g Saturated Fat: 0 g Cholesterol: 0 mg Sodium: 200 mg Potassium: 300 mg

13:

ZAP-PROOF MINERALS AND SUPPLEMENTS

One of the best-kept secrets in nutrition has been under wraps for over seventy years. And here it is: minerals are the spark of life and even more important than vitamins when it comes to cellular health. They are really the front line of defense in protecting the body against increasingly pervasive EMFs.

Think of it this way: Every cell relies upon minerals for numerous bodily functions. And there are over seventy trillion cells in the human body, each one acting like a biological battery—a kind of mini-dynamo that generates life. Minerals are the catalysts to keep the battery going, help it hold a charge. Without minerals in the proper ratio, cellular membranes can't maintain the proper liquid pressure between the inside and outside cell walls. It's called osmotic pressure, and it's what keeps viruses and bacteria from invading and cells from rupturing. Those same minerals, usually called electrolytes, help maintain the right pH in intracellular fluid so alien organisms that do slip by won't survive. Any imbalance in this pressure causes cells to weaken and eventually die. Even the immune system depends upon this mineral balancing act, right down to the cellular level. The minerals of life recharge us on a minute-by-minute basis, empowering every cell, organ, and tissue of the body.

Since EMFs target your cells—the membranes in particular—you must keep your minerals at top notch levels 24/7 to protect and defend yourself from cell collapse. We already know the stealth stress generated by EMFs creates free radical damage. But it does something more: like emotional stress, it floods the body with stress hormones, which produce more and more chemicals to transmit nerve impulses. Remember, when you're under stress, your body goes into emergency mode so it's sending out messages as frantically as a military platoon under siege. Most of those chemicals are electrolytes or trace minerals, which your body needs to produce antioxidants like superoxide dismutase (SOD) to help scavenge those free radicals. But if you're under chronic stress—and believe me, EMFs are with you nearly all the time—you rapidly use up your mineral stores. Studies have found that it can take a week or more to restore those minerals to the body when the stress subsides. When you're under unremitting stress, you need to restore trace minerals *on a daily basis.*

Minerals do much more in addition to their role in protecting the cell membrane against electropollutants. They promote blood formation, fluid regulation, protein metabolism, and energy production. They assist in every biochemical response in the body. Minerals help maintain strong bones and teeth, provide organ and glandular strength, and relieve stress. Since minerals are obtained through the soil, recognizing their value is one thing; getting them into your system is quite another matter altogether.

WHERE HAVE ALL THE MINERALS GONE?

Simply put, you can't get minerals from depleted soil, or from processed or overcooked foods. For years nutritionists have recommended that everyone eat leafy greens because they contain many nutrients, including chlorophyll and calcium.

For this reason, it will be extremely helpful for you to have a baseline study of your body's mineral levels, which is most accurately and non-invasively done with a tissue mineral analysis or hair test. It provides a unique window into individual mineral levels and key mineral ratios (as well as heavy metals, which can displace minerals) in the tissues. The hair provides a two- to three-month blueprint of what has occurred, metabolically speaking, in the body. If you discover that you have deficiencies or excesses or simply want to shore up your EMF shield, zero in on the minerals listed here and the foods that contain them.

Calcium

You know calcium is vital for strong bones and teeth, but since EMFs can dramatically alter the calcium content of your cells, absorbing enough of this macro mineral is vital. Calcium is the mortar that holds your cells together. EMFs break through the membrane and calcium leaks out, which can cause all kinds of symptoms large and small. Strive for at least a 1:1 ratio with magnesium, calcium's sister mineral.

Optimum dose: Maximum of 800–1000 mg per day for all ages (including children) provided there is sufficient magnesium and vitamin D intake via foods and/or supplements. If taking supplements, this is best taken in 400–500 mg dosages. My personal favorite is Osteo-Key (see resources) which contains hydroxyapatite, providing one of the most bioavailable sources of calcium along with magnesium.

Good food sources: Dairy foods; dark leafy greens; almonds; pecans; sunflower seeds; parsley; kelp; burdock; and hijiki.

Magnesium

Magnesium is your best insurance policy for both absorbing and retaining enough calcium. Magnesium keeps a check on calcium in your nerve cells to prevent overstimulation. A keystone antistress mineral, it also plays a major role in protecting your arteries, nerves, muscles, and bones and keeps your blood circulating smoothly.

Optimum dose: 400–800 mg for most adults. If loose bowels occur, cut back on dosage. 50–400 mg for children and teens.

Good food sources: Green leafy vegetables; broccoli; nuts, especially almonds; seeds, especially pumpkin, sunflower, and sesame; and beans.

Increase Melatonin Naturally

Supplements and food aren't the only way to increase your body's production of melatonin. Adopting a few new lifestyle habits will also help.

Get some sun. I know, it sounds contradictory since sunlight suppresses melatonin production. But in studies of elderly people, who tend to have lower than normal levels of melatonin and consequently have sleep problems, researchers have found that those who were exposed to about two hours of sunlight a day had higher levels of melatonin—and slept better—than those who didn't spend as much time in the sun.[1]

Get some artificial sun. If you live in a place where sunlight is in short supply, particularly in the winter, consider purchasing full spectrum indoor lighting, which mimics the sun. Long-recommended for treatment of the winter blues—seasonal affective disorder, or SAD—full spectrum light can help you maintain regular circadian rhythm during those literal dark times and help keep melatonin flowing at night.

Keep it dark after dark. Lux is a measurement of how brightly surfaces are illuminated by light. On an average day, sunlight ranges from about 32,000 to 100,000 lux. Average indoor lighting ranges from 100 to 1,000 lux. The full moon is about 1 lux, and a movie theater is less than 1 lux. A dark, overcast day is less than 1 lux. Studies have found that even exposure to as little as 100 lux at night can inhibit melatonin production. At 500 lux, light exposure has impaired melatonin in humans in the range of up to 98 percent. So keep the lights low or even off in the hours leading up to bedtime.[2]

Phosphorus

You'll find phosphorous in most foods because it's vital to every living thing. Your cell membranes are constructed, in part, of phospholipids. All energy production and storage is dependent on compounds that include phosphorous (such as adenosine triphosphate, or ATP) and nucleic

Darken up. Our world tends to be over-illuminated at night. Light from cities even obscures the stars. In fact, if you live near an urban center, your sky can look like a not-so-pretty aurora borealis in the evening. If there were no artificial lights, we'd be producing melatonin nine to ten hours a night. The average American only gets about seven to eight hours of darkness every night, and may only secrete melatonin for six to seven of those. So not only does that constant glow waste an estimated one hundred thirty billion dollars a year in energy costs, we pay a health price too. As nature writer Verlyn Klinkenborg wrote in his article "Our Vanishing Night" in *National Geographic* magazine in November 2008, "Darkness is as essential to our biological welfare, to our internal clockwork, as light itself."[3] To protect yourself, consider blackout shades to shut out your neighbors' lights.

Eschew blue at night. On the light spectrum, blue light is great for helping you stay alert. One reason: studies show it suppresses melatonin production. Since we're dealing with an invisible force that does the same thing, it's important to filter out blue light to keep melatonin production at its peak. Researchers studying night-shift workers, who are at high risk for breast cancer and sleep disorders, found they could prevent melatonin suppression by outfitting the workers with glasses that blocked the artificial blue light on the job. The workers slept better and may have been less vulnerable to cancer.[4] Researchers at the Lighting Innovations Institute at John Carroll University in Ohio developed reasonably priced blue-blocking glasses, light bulbs, and night-lights that their studies found helped promote better sleep and reduced symptoms of ADHD in people who wore them a few hours before bedtime (see resources).

To a large extent, we have the power to protect ourselves and our children from the unchecked and unheralded effects of electropollution. Highly mineralized foods, dietary supplements, and all-star protectors will maximize our body's defenses and safeguard our well-being.

acid—the *NA* part of your DNA and RNA, which store and communicate your cellular genetic information and are made up of strings of molecules containing phosphorous. Cellular communication would be impossible without this important mineral.

Optimum dosage: 700 mg for adults nineteen and older; 700–1,250 mg for pregnant or lactating women; 1,250 mg for children nine to eighteen.

Good food sources: Wheat bran; herring; sesame seeds, cashews, liver, all-bran cereal, egg yolk.

Potassium

This is one of your body's key electrolytes—along with sodium, chloride, and bicarbonate—so called because they conduct electricity in water. Potassium is a positively charged ion concentrated in the fluid inside the cell membrane. Along with positively charged sodium on the outside of the membrane, potassium helps create what's known as the membrane potential—the voltage across the cell membrane that keeps it stable and also helps your cells communicate with one another.

Optimum dosage: The best way to get enough potassium is through food.

Good food sources: Coconut water; bananas; potatoes; tomatoes; Swiss chard; lima beans; yam and winter squash; soybeans; avocado; spinach; pinto beans; papaya; lentils.

Sulfur

One of the nonmetallic elements, sulfur plays a starring role in the anti-oxidant power of glutathione. Glutathione's sulfur atoms latch on to toxic molecules and free radicals and either make them easier to eliminate or turn them into harmless compounds. Since EMFs can reduce your body's levels of glutathione, make sure you get enough of this building block to keep it at optimal levels.

Optimum dosage: There is no official recommended amount. However, supplements like MSM provide sulfur.

Good food sources: Cruciferous vegetables; cabbage; onions and garlic; meats, fish, legumes, and eggs.

Chromium

High blood sugar has been noted so often in studies of EMF exposure that one researcher refers to the EMF-triggered phenomenon as "type 3 diabetes."[5] Chromium helps by doing what it does best—helping insulin turn blood sugar into energy instead of leaving it circulating in the blood. Chromium is also directly involved in protein, carbohydrate, and fat metabolism, hence its reputation as a fat-burner and potential weight loss aid.

Optimum dosage: Adults aged nineteen and older: 200–400 mcg a day; pregnant or lactating women: 200 mcg a day; children fourteen to eighteen: 35 mcg for boys, 24 mcg for girls.

Good food sources: Brewer's yeast, oysters; liver; whole grains; potatoes; raw onions; romaine lettuce; ripe tomato.

Manganese

While manganese helps your body metabolize fat and protein and build strong bones, I recommend it because it is one of the underrated minerals so absolutely critical to form the powerful SOD, which is inside your mitochondria (the energy powerhouse of every cell). SOD, as you may recall from previous chapters, is one of the key bodyguard enzymes that protects us from cell-damaging free radicals created by even low-level EMF fields. It also helps to produce the thyroid hormone thyroxine, which helps to protect your thyroid from EMF damage.

Optimum dosage: Adults, including pregnant and lactating women: 15–30 mg; children fourteen to eighteen: 9 mg.

Good food sources: Leafy greens such as kale, chard, romaine lettuce, and turnip greens; fruits such as strawberries, figs, kiwi, and pineapple; whole grains; legumes; garlic; herbs and spices such as cinnamon and turmeric.

Selenium

This trace mineral is a potent heavy metal chelator (especially of mercury) and antioxidant, vital to the body's synthesis of glutathione, the body's premier detoxifier. Lack of selenium magnifies the effects of io-dine deficiency in thyroid disorders. Selenium enhances the immune system, especially in immune-suppressed individuals or those who are electrosensitive. It has also been identified as a cancer fighter, particularly against lung, colorectal, prostate, and non-melanoma skin cancers. Studies have found it works in many ways beyond its role as a free radical scavenger: it directly inhibits tumor growth and suppresses the development of blood vessels that feed tumors. Please note that selenium is included in many dietary supplements, and some people do have excessively high selenium levels, which can lead to nerve damage, osteopenia, as well as liver and kidney impairment. On the other hand, selenium is deficient in most soils in the United States, with the exception of the Dakotas, so do consider a trace mineral analysis to determine your own specific needs with this mineral in particular (see resources).

Optimum dosage: Adults and children over fourteen: 100–200 mcg.

Good food sources: Brazil nuts; beef, poultry, turkey; fish such as snapper, cod, shrimp; button mushrooms, shiitake mushrooms; barley.

Zinc

Zinc is essential to the functioning of every single cell in your body; it plays a role in about a hundred enzyme reactions. In one study, animals exposed daily to EMFs and then given zinc had higher levels of glutathione than animals similarly exposed but who weren't given the mineral.[6] Zinc also helps with DNA synthesis, cell division, wound healing, and overall immune support.

Optimum dosage: Adults nineteen and over: 50 mg; pregnant and lactating women: 50–100 mg; children fourteen to eighteen: 25 mg.

Good food sources: Calf's liver, venison , meat, poultry; legumes; mushrooms; kelp, spinach; pumpkin seeds; and eggs.

ALL-STAR EMF PROTECTORS

You can outsmart, and hopefully reverse, any potential health crisis in the making. Arm yourself with this research-based arsenal of specific zap-proof supplements. Shield yourself now!

If you are one of the sensitive individuals who cannot tolerate dietary supplements—as crucial as they are—you can use Trace-Lyte Minerals, a homeopathic form of organic minerals that strengthen the cell walls to protect against electropollution and other environmental toxins (see resources).

Melatonin

If you decide to have your blood levels of melatonin checked, make sure you do it at night. Since melatonin is what puts you to sleep, it's naturally lower during the day and shifts upward at night. Melatonin production peaks between two and four A.M., which is why making sure your bedroom is an EMF-free zone is so important.

Studies are pretty clear: Electropollution can seriously reduce the body's production of this vital antioxidant-like hormone, which has been linked to cancer protection and which also plays a role in regulating the body's two most powerful free radical scavengers—glutathione and SOD, both liver enzymes involved in the body's detoxification system. Melatonin, unlike other antioxidants, is able to cross the blood-brain barrier—meaning that it can actually breach the membrane designed to prevent toxins from entering your brain[7]—and has been shown to be twice as effective at preventing free radical damage to cell membranes (one of the prime targets of EMFs) as vitamin E. It's been found to be five times more effective than glutathione at neutralizing hydroxyl radicals, which may be responsible for more than half of all free radical damage in the body, and also targets the superoxide radicals.

I recommend that everyone, but especially those over sixty, take melatonin, which substantially decreases as we age. This is definitely one supplement you will want to take.

Optimum dosage: 1–3 mg, preferably in a time-released form.

Glutathione

One of the reasons glutathione is such a potent free radical scavenger is location—it's in every cell of your body, giving each cell its own personal bodyguard. It protects your DNA, keeps you energized, reduces inflammation and pain, and neutralizes toxins and heavy metals. It is one of four substances absolutely necessary for cell survival, but you need to work to keep your stores intact: glutathione levels drop by about 10 to 15 percent every decade of life. If you're sick, stressed, or toxic and take lots of over-the-counter or prescribed medications, your glutathione levels can nosedive. Deficiencies are frequently seen in those who have cancer, Alzheimer's or Parkinson's diseases, or immune-deficiency conditions like AIDS. But if you're exposed to EMFs—and we all are—it appears to be crucial.

Glutathione isn't absorbed readily when taken orally. It's digested before it even reaches your bloodstream. A glutathione intravenous push or suppositories (made at compounding pharmacies) are some of the better delivery systems, both of which require a doctor's supervision. Your best bet is to make sure you eat right (broccoli sprouts are one of the highest promoters of your body's glutathione production) and take supplements that boost your body's own natural production of this major league cell protector. They include:

NAC (N-acetyl-L-cysteine)

Studies have shown that NAC, derived from the amino acid L-cysteine, will help raise levels of glutathione in the cells. A sulfur compound, it also acts as an antioxidant on its own. There's also evidence that it boosts the immune system. In one study, Italian researchers created a scientific cage match between staph germs and immune cells. When they added NAC to some of the cultures, the germ-fighting ability of the cells was significantly enhanced. Not only that, they literally lived to fight another day: many immune cells die in the process of subduing bacteria, but more of the NAC-treated cells survived. NAC has also been studied as a cancer fighter. In that battle, it seems to prevent DNA damage by blocking the effects of carcinogenic compounds on cellular genetic material and by quelling the free radicals produced by cancer cells, which signal other cancer cells to grow.

Optimum dosage: Up to 600 mg (see resources).

Vitamin D

Vitamin D has yet another claim to fame—as if building bones and fighting depression, dementia, heart disease, and flu are not enough. Calcitrol, the active form of vitamin D3, is an ideal radioprotectant even against low-level background radiation.[8] It helps facilitate communication between cells, which is interrupted by EMFs, and also activates the immune response, which can be tamped down by exposure to radiation.

Of course, the exact amount of vitamin D you need to optimize blood levels (50–80 ng/ml) varies considerably based on your age, genetics, how far north you live, the season of the year, and how much time you spend in the sun. The 25-hydroxy vitamin D blood test is the best way to figure out your current levels, and then you can adjust your dose accordingly.

Optimum dosage: 5,000–10,000 IUs of vitamin D3.

Whey Protein Powder

Whey protein is one of the richest sources of the amino acids from which glutathione is derived. It also contains naturally occurring healing substances such as lactoferrin, an iron-modulating protein; immunoglobulins, immune-enhancing elements; and glycomacropeptides, natural appetite-suppressing proteins.

Look for the highest quality whey protein concentrate derived from herds that are only fed chemical- and pesticide-free grass. The whey protein concentrate should also be unheated and non-denatured to keep the glutathione-building amino acids intact. Make sure the product is lactose-free powder with no added sugars or artificial sweeteners. Use as the base of a smoothie on a daily basis. You can even increase its protective health properties if you mix in some of the Zapped Superfood fruits, such as blueberries or frozen açaí berries.

Optimum dosage: 1–2 scoops containing 20 grams of protein per scoop.

Ultra H-3

First discovered in 1949 by Dr. Ana Aslan of Romania's National Geriatric Institute, procaine HCl repairs damaged cell membranes so they can absorb nutrients more effectively. More than five hundred laboratory studies by leading researchers show that procaine helps protect the brain from damaging electrophysiological changes. Its extraordinary rejuvenating powers are due to procaine's ability to balance monoamine oxidase for improved mental clarity and to enhance circulation to the brain.

A patented procaine product, Ultra H-3—in a matrix of ascorbic acid, citric acid, niacin, folic acid, biotin, and magnesium—lasts about fifteen times longer than Aslan's original discovery and is considered 100 percent bioavailable. Ultra H-3 also contains gingko and bilberry extracts to ensure targeted delivery of nutrients through the blood-brain barrier.

Optimum dosage: 1 to 2 tablets twice daily with a glass of water, 6–8 hours apart, 1 hour before or 2 hours after eating, or as directed by a health care professional (see resources).

Milk Thistle

Milk thistle contains the phytochemical silymarin, which not only increases levels of glutathione in the body by as much as 35 percent but also boosts the other EMF-targeted antioxidant, SOD.[9] In addition, it aids in cell repair, calms inflammation, and helps detox the liver (it's actually used to treat cirrhosis and nonalcoholic liver disease as well as acetaminophen and mushroom poisoning). Numerous studies have found that silymarin reduces oxidative stress.

Optimum dosage: 140 mg of a standardized extract twice daily.

Alpha Lipoic Acid

An effective free radical scavenger, alpha lipoic acid (ALA) earns its stripes by increasing glutathione and giving new life to other antioxidants in your diet. Most free radical scavengers like vitamins C and E and glutathione, when they latch on to unpaired electrons, eventually become free radicals themselves. ALA reduces these oxidized forms of antioxidants (including another, Coenzyme Q10, which plays a role in your cell's mini engines, the mitochondria) and makes them come to life again. It also seems to improve the role of insulin—helping this pancreatic hormone usher sugar from the blood to your cells, where it's burned for energy— to slow age-related cognitive decline, and to reduce the symptoms of diabetic neuropathy, damage to the nerves caused by high blood sugar. CAUTION: *Start with the lowest dose if low blood sugar is a concern. This supplement reduces blood sugar dramatically—and quickly.*

Optimum dosage: 200–600 mg

Superoxide Dismutase

SOD supplements have been used to reduce tissue injury brought on by radiation treatments.[10] Studies show it acts not only as an antioxidant—it quenches superoxide radicals and repairs cellular free radical damage— but also as an anti-inflammatory in the body. It also helps the body use the trace minerals zinc, copper, and manganese, which defend the mitochondria from free radical damage. Studies have also found a strong link between occupational EMF exposure and ALS. Although there's no leading theory on a possible biological explanation, European researchers who found changes in "antioxidant defense systems" in the brain think that could provide a clue.

Optimum dosage: 5500 units, 1–3 times per day.

OTHER ZAP-PROOF SUPPLEMENTS TO CONSIDER

These nutrients include some frequently overlooked yet powerful antioxidants and other elements that have been shown to be effective in protecting against the side effects of both ionizing and nonionizing radiation exposure, including EMFs.

Coenzyme Q10 (ubiquinol)

Under ideal conditions, Coenzyme Q10 (CoQ10) is the premier heart-protecting compound naturally synthesized by your body. It acts as an antioxidant in your cell membranes, protecting your DNA and lipoproteins, where it prevents the oxidation of blood fats, which makes them less dangerous to your arteries and cardiovascular system. It's absolutely vital to your mitochondria—without it, your cells aren't able to convert carbohydrates and fats into the fuel they need. Your body doesn't make as much CoQ10 as you age, so adding this supplement to your daily regimen is a good idea. It's vital, especially if you take a statin drug for lowering cholesterol, as statins (like Lipitor and Zocor) dramatically reduce the amount of CoQ10 your body produces.

Optimum dosage: 100–300 mg.

Honeybee Propolis/Royal Jelly

Propolis—the resin bees make to build their hives to protect against pathogens and molds—was the source of caffeic acid used in animal studies showing protection from free radical damage when rats where exposed to cell phone EMFs (900 MHz) for thirty minutes a day for ten days. The animals that weren't given the propolis experienced a halt in SOD and glutathione production, particularly in their hearts.[11] Other studies have found that caffeic acid, which is also found in coffee, cruciferous vegetables, citrus fruits, apples, and pears, protected lymphocytes (immune system cells) from gamma radiation damage. It also acts as an anti-inflammatory and protects against cancer. I especially like Dr. Ohhira's Propolis Plus from Essential Formulas because this product also contains healthy probiotics for intestinal fortification, where 75 percent of immune receptor sites are located (see resources).

Optimum dosage: 100–500 mg.

Sea Buckthorn

The leaves and berries from the sea buckthorn deciduous shrub found in the East contain antioxidants galore: beta carotene, flavonoids, polyphenols, vitamin E, and even lycopene, along with a healthy dose of trace minerals calcium, potassium, and magnesium, and essential fatty acids. In studies, sea buckthorn prevented the oxidation of cholesterol—a process that makes it more likely to stick to artery walls—and reduced C-reactive protein levels, a marker of inflammation linked to heart disease.[12] You can consume sea buckthorn as a juice or tea.

Optimum dosage: 500 mg

EPILOGUE

Zapped is just the beginning of a global awakening to the challenges presented by electropollution.

Clearly, we are only starting to grasp the consequences of our love affair with clever digital gadgets and gizmos to make our lives more convenient or to entertain ourselves. Based on what we know today, this book offers some simple (as well as more extensive) solutions to the rising tide of electropollution, especially with regard to wireless technology. Here are a number of actions you can take immediately to minimize your EMF exposure:

- Don't use your cell phone as an alarm clock at night. Consider the old-fashioned battery alarm clock instead.

- During the day, keep your cell phone away from your body as much as possible. Use the speaker mode when talking, and don't carry your cell phone in your bra, back pocket, or on your person.

- Replace your cordless phones with a corded landline phone. Even when not in use, the charger is constantly emitting radiation.

- Turn off your wireless router whenever it is not in use and definitely at night. It radiates for nearly six hundred feet.

- Keep your laptop off your lap. Who wants to radiate their reproductive organs?

- Unplug electrical appliances when not in use. They are still emitting electricity.

- Keep electronic equipment out of your bedroom. Remember, you need to have a safe haven to regenerate, especially during sleep.

- Only have CT scans and X-rays when absolutely necessary. This kind of radiation is cumulative.

- Keep your babies and children safe by removing *all* electronic games, equipment, and monitors from their bedrooms. Remember, little human beings, due to their size and thinner skulls, are way more vulnerable to EMFs than adults are.

The electrosensitive individuals profiled in this book give us a glimpse into what can eventually happen to us all—especially children—if we don't act now. Being proactive by choosing which electromagnetic devices you really can live without, which ones you absolutely need, and how to use them safely is certainly a smart move in the right direction. You may also want to support your body with the right diet, supplements, and grounding devices, which help to discharge the accumulation of bioelectrical stress.

Simply by becoming aware of the problem, you've taken the critical first step toward solving it. You don't have to become a full-time political activist to do your part in spreading awareness about this important issue, but I do recommend that you become familiar with the organizations that are on the front lines in this battle and support them in any way you can—through donations, signing of petitions, volunteer work, and telling others about them.

There are a growing number of environmental consultants and inspectors who include assessment of the electromagnetic environment as part of their services. The International Institute for Bau-Biology ["building biology"] & Ecology is an organization that trains people in this regard. Bau-Biology–trained consultants and inspectors throughout the country have the know-how to identify and rectify a variety of stress factors in your indoor environment, including electromagnetic pollution. If you choose to do your own measurements of electromagnetic fields—as detailed in the book—they, or an alternate EMF remediation source, can still be very insightful.

Blanketing much larger areas, long-distance Wi-Fi or Wi-Max—the multibillion dollar wireless network involving Clearwire, Google, Sprint, and Time-Warner—will undoubtedly affect even more people in the future. Consumers really need to ask Congress for premarket health testing before allowing the rapid expansion of this kind of "Wi-Fi on steroids."

We have the power to change our environment to make it more livable and sustainable—not only in our own personal lives but in our communities. So, even more important will be demanding Congress to take action to fund independent EMF research and repeal Section 704 of the Telecommunications Act, so that state and local governments no longer have their hands tied in trying to protect human health from unwanted exposure.

Who knows what new sources of electropollution will arise in the future?

According to *The Economist,* "already there's talk of encouraging consumers to install 'femtocells,' wireless base stations in their homes. While it clearly has a vested interest in mobile-data traffic, Cisco expects Wi-Fi to increase 39-fold over the next five years."[1]

As you have seen—and I will say it again—the younger and smaller an individual is, the more at risk. All parents—not to mention aunts, uncles, and grandparents—want to protect future generations, including those yet to be born. Hopefully, with the information provided in this book, you will have a better sense about the level of electropollution to which you are willing to expose yourself—and your children.

While this book can only address the potential dangers we know about today, what it does hope to offer is a basic understanding of the growing risks of EMFs. They may be invisible, but they are very real. No doubt they will increasingly continue to challenge everybody's "body electric."

So go now and make the right choices. The resources section that follows is designed to help you simplify your search for the most protective supplements, detection devices, and the best educational resources for further study and updates. If you do your homework now, I predict that not only will you will be blessed with better health, but your home and office will be better protected, and you will be helping to create a more sustainable Mother Earth.

"All truth passes through three stages.
First, it is ridiculed. Second, it is violently opposed.
Third, it is accepted as being self-evident."

—*Arthur Schopenhauer*

ZAPPED RESOURCES, SUPPORT, AND SOLUTIONS

ONLINE SUPPORT

The Official Zapped Web Site:
www.areyouzapped.com
Check in daily for the most persuasive articles, pertinent updates, and breaking news regarding electropollution. Here's where you'll also find EMF and grounding products, meters, protective devices, and supplements to thrive and survive in the digital age.

Dr. Ann Louise's Edge on Health Blog
www.annlouise.com/blog
Several times a week, get my take on cutting-edge news and information on the hottest topics in health. Look for special "Zapped" blogs so that you can be informed as soon as I am.

EDUCATIONAL WEBSITES

American Academy of Environmental Medicine
www.aaemonline.org
The American Academy of Environmental Medicine is an international society of physicians and other professionals studying the clinical aspects of human health as impacted by the environment. The AAEM

promotes optimal wellness by recognizing, working to prevent, and safely treating environmental illnesses and sensitivities.

BioInitiative Working Group
www.bioinitiative.org
The BioInitiative Working Group is an international group of researchers, scientists, and health policy officials. In 2007, the group released a report on electromagnetic fields and health, documenting serious public health concerns about current supposed allowable limits regarding EMFs from cell phones, power lines, and many other sources of exposure in everyday life.

Campaign for Radiation Free Schools (Facebook Group)
www.facebook.com/group.php?gid=110896245588878
A resource for schools, teachers, and parents, including EMF audio interviews with experts, news, and the EMF Help Blog™. Encourages schools to "BRAG Grade" themselves following the risk assessment protocol of the BRAG Antenna Ranking of Schools Report by Magda Havas, Ph.D.

International EMF Alliance
http://international-emf-alliance.org/
New collaboration of the world's EMF advocacy organizations on cell phones and wireless and other EMF hazards that is coordinating closely with the world's leading EMF scientists to disseminate vital scientific information to governments, the medical profession, and consumers.

International Institute for Bau-Biology & Ecology
www.buildingbiology.net
The International Institute for Bau-Biology & Ecology is a nonprofit North American organization that combines building biology, ecological principles, and technical expertise to support healthy living and a more sustainable environment, according to the precautionary principle.

Centers for Disease Control and Prevention

www.cdc.gov/niosh/topics/EMF/

The Centers for Disease Control and Prevention maintain this site on potential EMF hazards.

Dirty Electricity

www.dirtyelectricity.org

Dirty electricity—the name says it all! However, this phenomenon, as you have learned, is not widely known and can be complex to understand. This site provides information on it.

The Earthing Institute

www.earthinginstitute.net

All the latest scientific studies proving the efficacy of Earthing can be found at this site.

Electrical Pollution Solutions

www.electricalpollution.com

This site is managed by Catherine Kleiber, who is featured in this book, and provides good basic information on the politics and science behind electropollution.

Electromagnetic Health

www.electromagnetichealth.org

Created by EMF activist Camilla Rees, this website provides informative audio clips and videos from some of the world's leading EMF experts. To protect your family and your own health, sign the petition posted on this site. It's important that Congress:

1. Require the FCC to revise RF exposure guidelines.

2. Repeal Section 704 of the Telecommunications Act of 1996 that makes state and local governments powerless to stop cell towers and wireless antennas based on "environmental [i.e., human health]" concerns.

3. Stop further wireless infrastructure development until more research can be done.

4. Establish wireless-free areas in every community, particularly in public building and transportation options.

Energy Medicine Foundation
www.energymedicinefoundation.org
Carol Keppler's Energy Medicine Foundation serves as an advocate for the acceptance of energy medicine through research and education while funding, sponsoring, designing, and promoting primary research in cutting-edge energy medicine advances and related modalities.

Environmental Health Trust
www.environmentalhealthtrust.org
Devra Lee Davis, Ph.D., M.P.H., is a visiting professor at Georgetown University and is the developer of the Environmental Health Trust.

Guinea Pigs "R" Us
www.guineapigsrus.org
This educational website highlights risks from long-term EMF/EMR, particularly nighttime exposure and its links to ADD/ADHD, Alzheimer's disease, autism, leukemia, and more.

Lisa Nagy
www.lisanagy.com
Lisa Nagy, M.D., herself a victim of multiple chemical and electromagnetic sensitivities, operates an informative website that contains helpful suggestions for taking your own environmental health history.

Microwave News
www.microwavenews.com
Operated by Louis Slesin, Ph.D., a former scientist for the Natural Resources Defense Council, this website is "meticulously researched and thoroughly documented," according to *Time Magazine*. Slesin also offers a free e-mail newsletter that will keep you up-to-date on new findings.

Power Line Health Facts

www.powerlinefacts.com

The website is maintained by the Power Line Task Force (PLTF), a group of proactive homeowners living in Minnesota.

Powerwatch

www.powerwatch.org.uk

Powerwatch is a nonprofit independent organization in the United Kingdom that has been researching the effects of EMFs on health for more than twenty years. It also works with decision makers in government and business and provides a wide range of educational resources for the lay public.

EM Radiation Research Trust

www.radiationresearch.org

The EM Radiation Research Trust is an independent organization whose mission is to provide the facts about electromagnetic radiation and health to the public and the media.

The World Health Organization

www.who.int/peh-emf/en/

In 1996, the World Health Organization (WHO) established the International EMF Project to assess the scientific evidence of possible health effects of EMF in the frequency range from 0 to 300 GHz. This vast site contains fact sheets and reports on the WHO findings.

EMF TECHNOLOGIES AND PRODUCTS

Less EMF Inc.

809 Madison Avenue

Albany, NY 12208

888–537–7363

518–432–1550

www.lessemf.com

Less EMF Inc. provides a wide range of products—including EF and RF meters, Gauss meters, electrosmog detectors, and shielding devices—for identifying and protecting against electromagnetic pollution.

emfsafetystore.com

For people who want to learn how to assess home and office environments for electromagnetic fields, this site aims to teach how to clean up your personal environment, remediating EMF safety problems, if they exist, especially sleeping and office environments where you spend most of your time. To accomplish that, they offer a selection of quality, value-added, and economical measurement and safety resources, screened by expert EMF consultant Stan Hartman of Radsafe in Boulder, Colorado.

LowBlueLights.com

7890 Summerset Drive
Walton Hills, OH 44146
216–397–1657
www.lowbluelights.com
LowBlueLights.com carries special glasses and night-lights that block out blue lights that inhibit melatonin production. I own a pair of these glasses and wear them an hour before bed, which helps me sleep through the night without waking.

ONDAMED Inc

2570 Route 9W
Cornwall, NY 12518
800–807–7864
845–534–0456
www.ondamed.net
ONDAMED—which Stephen Sinatra, M.D., calls the "future of medicine"—is a battery-powered device that combines biofeedback with pulsed electromagnetic fields to enhance electrical potential of weak tissue, promote relaxation, and relieve pain safely and noninvasively. This FDA-registered device has been shown by clinical

research to be effective in controlling Lyme disease, fibromyalgia, migraine, interstitial cystitis, and other conditions—without adverse effects.

Peninsula Quantum Wellness
Trevor May
P.O. Box 20005
Sidney, B.C. V8L 5C9
866–315–0588,
250–656–0588
www.quantumwellness.ca
Created after years of research and development by Professor Bill Nelson in Hungary, the EPFX-SCIO biofeedback device combines holistic healing with quantum technology to utilize the body's own natural electricity in detecting imbalances and treating numerous stress-related conditions.

QLaser Healing Light
520 Kansas City Street, Suite 100
Rapid City, South Dakota 57701
605–791–6060
www.qlaserhealinglight.com
QLaser Healing Light provides low-level laser therapy to reduce inflammation, increase blood circulation, stimulate bone repair, and relieve pain. FDA-cleared for osteoarthritis of the hand, cold laser therapy has been found safe and effective for a variety of conditions from asthma and dental problems to sciatica and tendonitis. I own one of these devices and find that it is helpful to accelerate healing. Dr. Larry Lytle's healing light has "the potential to change medicine," says Harry Whelan, M.D., neurology professor, Medicine College of Wisconsin.

VibesUP!
152 Cruickshank Drive
Folsom, CA 95630
916–984–9699
www.vibesup.com

Fun and whimsical, VibesUP! uses natural crystals and live extracts of plants to keep you grounded—quite literally. Much like grounding wires that run from electrical outlets in your home or office to the earth, VibesUP! Divine Mats reconnect—or ground—you to Earth's natural electrical energy—improving the quality of your sleep, relieving restless leg syndrome, and reducing soreness and stiffness. There are twenty unique products—from shoe inserts to teddy bears—to aid in neutralizing and releasing toxic energy.

ADDITIONAL INTERNAL EMF PROTECTION AND DETOXIFICATION

AltWaters Technology
1931 West Sweetwater Drive
Phoenix, AZ 85029
800–971–4458
www.altwaters.com
AltWaters Technology carries an entire six-step system of uniquely designed homeopathics that can reduce "electric overload" and "severe electric overload" with a unique series of maintenance formulas. The site also carries the Maltese Medallion, a type of protective sterling silver jewelry that is said to deflect harmful electric and magnetic influences.

Yarrow Environmental Solution
www.essencesonline.com/FES-YSF.htm
www.fesflowers.com/yarrow-formula.htm
Yarrow Environmental Solution (YES) is a healing combination of flower essences and whole plant extracts in sea-salt water to help protect against environmental toxins, including EMFs and other forms of radiation. YES carries a unique herbal and flower essence combination. A placebo-controlled study by Jeffrey Cram, Ph.D., of the Sierra Health Institute in California, tested the effects of YES's flower essence formulas—Yarrow Special Formula and Five-

Flower Formula—on people exposed to fluorescent lights and their unnatural frequencies. The placebo group did exhibit increased agitation—a sign of stress in the face of danger, while subjects who took one or the other of the flower essences exhibited no signs of agitation.[1]

In the 1930s, British physician Edward Bach, M.D., discovered healing properties of plants, creating the original 38 flower remedies that still bear his name, although newer plant essences have since been discovered.

EMF INSPECTORS AND PRACTITIONERS

InterNACHI

www.nachi.org/emfs.htm

The International Association of Certified Home Inspectors (InterNACHI) is a non-profit organization offering exams, standards of practice, code of ethics, and accreditation for electrical, EMF, and other home inspectors.

Gust Environmental

www.gustenviro.com

805–644–2008

Electrical engineer and certified mold remediator Lawrence Gust is an environmental consultant and electromagnetic radiation safety advisor, as well EMF speaker and teacher, who analyzes and remediates sick buildings and homes.

Sage EMF Design

www.silcom.com/~sage/emf/cindysage.html

805–969–0557

Owned by environmental researcher Cindy Sage, Sage EMF Design provides EMF remediation for private homes as well as consulting with architects, planners, and municipal and national agencies.

Environmental Health Center-Dallas

www.ehcd.com

214–368–4132

Founded by William J. Rea, M.D., the Environmental Health Center-Dallas, Texas, medically tests and treats human health problems involving sensitivity to a variety of pollutants such as air (indoor/outdoor), chemicals, dust, EMFs, molds, pollen, and many more health problems as they relate to the environment.

REMEDIATION EXPERTS

California

Peter Sierck
www.etanddt.com

Stephen Scott
www.EMFServices.com

Colorado

Stan Hartman
www.radsafe.net

Camilla Rees
www.ElectromagneticHealth.org

Florida

Will Spates
www.ietbuildinghealth.com

New Jersey

Sal LaDuca
www.emfrelief.com

New Mexico

Dan Stih
www.healthylivingspaces.com

Tennessee
Vicky Warren
www.wehliving.org

Texas
Jim Beal
www.emfinterface.com

Wisconsin
Dave Stetzer
www.stetzerelectric.com

Ontario, Canada
Kevin Byrne
www.dirtyelectricity.ca

Magda Havas, Ph.D.
www.magdahavas.com

Rob Metzinger
www.safelivingtechnologies.ca

Robert Steller
www.breathing-easy.net

RECOMMENDED BOOKS, MAGAZINES, AND NEWSLETTERS

The Body Electric: Electromagnetism and the Foundation of Life
Robert O. Becker, M.D., and Gary Selden
Harper Paperbacks, 1998
This book, by exploring the electrical nature of the body, set the stage for fully understanding the information contained in *Cross Currents,* as well as other books dealing with electromagnetic radiation.

Cell Phones: Invisible Hazards in the Wireless Age: An Insider's Alarming Discoveries about Cancer and Genetic Damage

Dr. George Carlo and Martin Schram

Carroll & Graf Publishers, 2001

No one knows this subject better than George Carlo, Ph.D., J.D., who was head of the Wireless Technology Research Project commissioned by the Cellular Telecommunications Industry Association in the mid-1990s. Although hired to prove the safety of cell phones, Carlo and his 200+ researchers found just the opposite after more than five years of study and an expenditure of $25 million. Since that time, Carlo has made it his mission to educate consumers on the dangers of cell phones and wireless technology.

Cell Towers: Wireless Convenience? or Environmental Hazard?

Edited by B. Blake Levitt

Safe Goods Publishing, 2001

Prominent researchers, government officials, engineers, and environmental attorneys contributed chapters to this book which explores the practicalities of cell towers.

Cross Currents: The Perils of Electropollution, The Promise of Electromedicine

Robert O. Becker, M.D.

Tarcher, 1990

Although this book is older, it is still is widely quoted and revered by others in the field, as well as lay readers, which is testimony to its quality and staying power. *Cross Currents* is the foundation—and sets the gold standard—for all who seek to educate others on this vital topic of electromagnetics.

Currents of Death: Power Lines, Computer Terminals, and the Attempt to Cover Up Their Threat to Your Health

Paul Brodeur

Simon & Schuster, 2000

Following court cases and scientific research, *New Yorker* writer Paul Brodeur explores risks ranging from cancer to miscarriage due to electromagnetic radiation.

Earthing, The Most Important Health Discovery Ever?

Clinton Ober, Stephen T. Sinatra, M.D., and Martin Zucker

Basic Health Publications, 2010

Earthing, the most eye-opening book to appear in decades, introduces the planet's powerful, amazing, and overlooked natural healing energy and how people anywhere can readily connect to it. The never-before-told story—filled with fascinating research and real-life testimonials—chronicles a discovery of the first magnitude with the potential to create a global health revolution.

Electromagnetic Fields: A Consumer's Guide to the Issues and How to Protect Ourselves

B. Blake Levitt

Backinprint.com, 2007

The former *New York Times* writer provides an exhaustive look at every aspect of electromagnetic fields, from naturally occurring fields to the plethora of diseases associated with electropollution. One of my personal favorites.

The Enemy Within: The High Cost of Living Near Nuclear Reactors

Jay M. Gould

Four Walls Eight Windows, 1996

This book reveals the high cost of living near nuclear reactors: breast cancer, AIDS, low birthweights, and other radiation-induced immune deficiency effects.

Energy Medicine: The Scientific Basis

James L. Oschman

Churchill Livingstone, 2000

If you've ever wondered how reiki treatments or massage therapy work, Oschman has the science.

Fat Flush for Life: The Year-Round Super Detox Plan to Boost Your Metabolism and Keep the Weight Off Permanently

Ann Louise Gittleman, Ph.D., CNS

Da Capo Press / Lifelong Books, 2010

One of the Top 10 Notable New Diet Books of 2010, according to *Time* magazine, *Fat Flush for Life* provides the latest generation of seasonal Fat Flushes accompanied by advanced detox and fitness regimens to increase metabolism while cleansing and strengthening the system for sustained weight loss, glowing health, and vitality. The book uncovers newly identified weight loss factors that target and correct imbalances in the GI tract and thyroid. A perfect detox and diet plan to follow year-round, which integrates many of the healing foods and supplements suggested in *Zapped*. The interactive messaging board at www.annlouiseforum.com provides round-the-clock support and motivation for Fat Flush followers.

The Fat Flush Plan

Ann Louise Gittleman, MS, CNS

McGraw-Hill, 2002

Also compatible with the *Zapped* dietary recommendations, the *New York Times* bestselling *The Fat Flush Plan* is the classic detox and diet that revolutionized weight loss in America. It offers a three-phase program that targets the liver and lymphatic system for melting fat from the tummy, hips, and thighs. This book provides easy and delicious menu plans and recipes with a fitness component. Online support is also available 24/7 at www.annlouiseforum.com.

Optimum Environments for Optimum Health & Creativity: Designing and Building a Healthy Home or Office

Dr. William J. Rea

Environmental Health Center-Dallas, 2002

Combining building expertise, medical research, and his own experience treating electrosensitive patients, William J. Rea, M.D., has written a valuable guide to creating environmentally healthy structures.

The Powerwatch Handbook:
Simple Ways to Make You and Your Family Safer
Alasdair and Jean Philips
Piatkus Books, 2006
For practical, easy-to-use tips on minimizing EMFs caused by a variety of popular devices and machines, this book shows how to reduce risks of potential health problems.

Public Health SOS: The Shadow Side of the Wireless Revolution
Camilla Rees, MBA, and Magda Havas, Ph.D.
CreateSpace, 2009
Written by a leading EMF activist and a prominent EMF scientist, this primer on EMFs and health grew out of 110 questions asked of the audience at the Commonwealth Club of California, a public affairs forum, in 2008.

Radiation Rescue: Safer Solutions for Cell Phones and Other Wireless Technologies
Kerry Crofton
Global Wellbeing Books, 2009
Drawing on the leading experts on the invisible hazards of electronic radiation, Kerry Croften sounds the alarm while providing science-based solutions for the whole family.

Silencing the Fields: A Practical Guide to Reducing AC Magnetic Fields
Ed Leeper
Symmetry Books, 2001
Physicist Ed Leeper, who was coauthor of the first study linking power line fields to leukemia in children, has written an easy-to-read (and actually amusing) guide to detecting and correcting electrical problems that lead to exposure to magnetic fields.

Tracing EMFs in Building Wiring and Grounding
Karl Riley

Magnetic Sciences, 2005

This step-by-step guide for both homeowners and professionals explains how to discover wiring problems that lead to EMF exposure as well as how to fix them.

Would You Put Your Head in a Microwave Oven?
Gerald Goldberg, M.D.

AuthorHouse, 2006

Combining biomedical research with a holistic mindset, internist Gerald Goldberg, M.D., explains how the human body is vulnerable to microwave radiation and offers ways we can better protect ourselves.

MAGAZINES

Taste for Life
100 Emerald Street, Suites A and D

Keene NH 03431

603–924–7271

www.tasteforlife.com

Taste for Life is the leading in-store magazine for health food stores, natural product chains, food co-ops, and supermarkets nationwide. Its excellent articles on pertinent health issues offer readers an informative educational source on a variety of levels including physical fitness. I sit on *Taste for Life*'s editorial board.

Total Health for Longevity
165 North 100 East, Suite 2

St. George, UT 84770–9963

888–316–6051

www.totalhealthmagazine.com

Total Health for Longevity is a comprehensive voice in antiaging, longevity, and self-managed natural health. It often features cutting-edge articles on electropollution. Publisher Lyle Hurd strives to bring readers fresh information and perspectives on all phases of longevity medicine so you can make an educated decision on the quality of your life today . . . and tomorrow.

NEWSLETTERS

Health Sciences Institute

Healthier News, LLC

702 Cathedral Street

Baltimore, MD 21201

888–213–0764

hsiresearch@healthiernews.com

www.hsibaltimore.com

As a member of the professional advisory panel, I can verify that this cutting-edge newsletter is devoted to presenting extraordinary products to its members before those products hit the marketplace. It was the first to break the Ultra H-3 story—the extraordinary product for arthritis, depression, and antiaging. The Health Sciences Institute provides private access to hidden cures, powerful discoveries, breakthrough treatments, and advances in modern, underground medicine.

Dr. Jonathan V. Wright's Nutrition & Healing

Healthier News, LLC

702 Cathedral Street

Baltimore, MD 21201

888–233–3402

www.wrightnewsletter.com

An alternative health newsletter that conventional doctors read, *Nutrition & Healing* explores the science behind prevention and treatment to prolong the quality of life.

Nutrition News

4108 Watkins Drive

Riverside, CA 92507

800-784-7550

www.nutritionnews.com

Siri Khalsa is a veteran journalist who has been in the business of providing health education for over twenty-five years. Her easy-to-read newsletter covers a wide variety of contemporary and current topics. It is distributed in health food stores throughout the country, but you can subscribe directly.

The Sinatra Health Report

7811 Montrose Road

Potomac, MD 20854

800-211-7643

www.drsinatra.com

Stephen Sinatra, M.D., FACN, CNS is a board-certified cardiologist and certified bioenergetic analyst with more than twenty years of experience in helping patients prevent and reverse heart disease. The Sinatra Health Report is published monthly by Healthy Directions, LLC. Sinatra is a big proponent of detoxification, and many of his newsletters discuss current research in the environmental medicine arena.

The Women's Health Letter

P.O. Box 467939

Atlanta, GA 31146-7939

800-728-2288

www.womenshealthletter.com

Nan Kathryn Fuchs, Ph.D., is the editor of this newsletter and is also my favorite nutritionist. Her comments regarding health, nutrition, and medicine as they relate to women are always on target.

OTHER HELPFUL RESOURCES

Hal Huggins, DDS, MS

www.hugginsappliedhealing.com

866-948-4638

Providing solutions for dental toxicity since 1973, Huggins has created an alliance of like-minded dentists who can provide healing solutions to all diseases linked to electrocharged materials in dentistry. His organization can test for biocompatible dental materials and make referrals to dentists throughout the country who follow the Huggins protocol for dental restoration. Huggins has been my personal dentist and dental adviser for decades.

Ranch Foods Direct

www.ranchfoodsdirect.com

866-866-MEAT (6328)

Ranch Foods Direct is a specialty meat company selling natural beef, raised without hormones or antibiotics, along with natural poultry, buffalo, eggs, cheese, pork, lamb, seafood, and many other quality food items. Combining a healthy meat-packing facility with naturally tender grass-fed beef and top-quality foods, rancher Mike Callicrate offers special discounts to readers using the code "ALG." Please visit his website to make your selection.

ACKNOWLEDGMENTS

There are many courageous researchers and scientists who are trying to sound the wired/wireless wakeup call but whose voices are rarely heard. I have sincerely tried to give many of their findings a platform in this book by presenting their research in an area charged with controversy and politics.

First and foremost, I must express my deepest thanks to the incredible dedication and research abilities of Denise Foley, who was invaluable in the development of this book. Denise embraced the project with joy, humor, and a scientist's desire to ferret out the truth. Denise: You are a gem. My brilliant and creative editor Nancy Hancock, who was also the guiding light in my Fat Flush series, once again showed her "true grit" with her tireless support and belief in the importance of the *Zapped* message. She engaged the entire HarperOne team, who have all been such enthusiastic supporters of this book—even as they work and live in the very high tech environment this book exposes. And, of course, many thanks to Coleen O'Shea, the editor of my first book and now a first-rate literary agent whose advice is always "spot on" and in the very best interests of all involved.

My gratitude goes out to my beloved mentor, the late (but still great) Dr. Hazel Parcells, who first opened my eyes to the electromagnetic energy of foods and the importance of cleansing the body of toxins, including radiation and all of its by-products. In a similar vein, I salute the late Robert O. Becker, M.D., and Richard Gerber, M.D., two pioneering researchers in the field of energetic medicine, whose books about energy medicine enthralled me. I am grateful for the work of biophysicist James Oschman, Ph.D.; Nobel Prize winner Gunther Blobel, M.D., Ph.D., who established that the cells of our bodies send out frequencies (energetic signals) when they need a specific nutrient. I stand in the greatest admiration of Samuel

Milham, M.D., M.P.H., of the Washington State Department of Health, who linked the rise in degenerative disease, cardiovascular disease, and suicide to U.S. electrification; the late W. Ross Adey, M.D., of the Pettis Memorial Veteran's Administration Hospital in Loma Linda, California, whose research discovered how electropollution disrupts cellular communication; George Carlo, Ph.D., J.D., who found nonthermal effects, including brain tumors and genetic damage, from cell phone radiation; Olle Johansson, Ph.D., associate professor of neuroscience at the Karolinska Institute, whose work showed how vulnerable children are to EMFs (and agreed—from Sweden—to be interviewed for this book); William Rea, M.D., the former surgeon who founded the Environmental Health Center in Dallas, Texas, to treat people with electrosensitivity; Martin Blank, Ph.D., associate professor of physiology and cellular biophysics at Columbia University, one of the world's leading researchers on nonionizing radiation; Nancy Wertheimer and Ed Leeper, whose groundbreaking research first linked childhood leukemia with power lines, and David Savitz, at the University of Colorado Medical School, who confirmed their findings.

I have been greatly inspired and enlightened by award-winning *New York Times* writer B. Blake Levitt, whose landmark work shows how damaging cell towers and EMFs are to human health; journalist Christopher Ketcham, whose *GQ* reporting showed how invested the telecommunications industry is in wireless (despite its health risks); and all the independent researchers and investigative reporters whose work continues to alert the public to the dangers of EMFs. I especially enjoyed the work of Kerry Crofton, Ph.D., whose latest edition of her landmark book *Radiation Rescue* will no doubt help safeguard millions.

My personal heartfelt gratitude is extended to Larry Gust, electrical engineer and certified building biologist, a renowned expert in identifying electropollution in the home, who discovered extreme EMF hazards in my own home. Thanks, Larry, for your meticulous attention to detail and for making my home truly zap-proof, which no doubt has helped to extend the quality of life for my family; Charles Keen, at EMF Services, for his insights in voltage mitigation; Clint Ober, a former cable TV pioneer, who developed personal grounding techniques to reduce the effects of EMF exposure, and his coauthors, Steven T. Sinatra, M.D., FACC, CNS, an esteemed colleague and respected cardiologist, and Martin Zucker, one of my favorite health journalists and dear friend, for their never-before-told story of natural healing energy in *Earthing*.

Kudos to Louis Slesin, Ph.D., whose *Microwave News* provides the latest updates and overviews in the world of microwaves and beyond. Thanks to Dave Stetzer, an industrial electrician, who with Martin Graham, Ph.D., professor emeritus in the college of engineering at the University of California at Berkeley, developed the Graham-Stetzer (G/S) filters, or Stetzerizers, that have helped thousands neutralize their "dirty electricity" and regain health; Magda Havas, Ph.D., advocate and associate professor of environmental and resources studies at Trent University, whose research found major improvements in blood sugar and stress levels with G/S filters. Dr. Havas was extremely gracious when interviewed for this book. More power to the work of Camilla Rees, activist extraordinaire, and Dr. Havas's coauthor of *Public Health SOS: The Shadow Side of the Wireless Revolution*.

I have the greatest admiration for the new breed of impassioned electrosensitive "canaries" that willingly shared their personal stories so that others could benefit. My thanks to Catherine Kleiber; Dr. Lisa Nagy, another victim of multiple chemical and electromagnetic sensitivities; and Gilligan Joy. There are myriad EMF experts and activists like Cindy Sage and David Carpenter, M.D. (coeditors of the landmark *BioInitiative Report*); Ann Baldwin, Ph.D., and Melinda Connor, Ph.D., at the Laboratory for Advances in Consciousness and Health at the University of Arizona; and Martha Howard, M.D., medical director of Wellness Associates of Chicago, who are doing exemplary work in the field. A very fond personal thanks to the myriad e-mails and support from the Energy Medicine Foundation's Carol Keppler, who has faithfully kept me abreast of the latest frequency updates with her special brand of TLC.

On the home front, the solid editorial/research/recipe assistance provided by Suzin Stockton; Gregg Stebben; Roon Frost; Bernie Rosen, Ph.D.; Linda Shapiro; Stuart Gittleman; Carol Templeton; and Tami Oliver (my personal assistant, who listened as I read and reread scientific jargon) were grounding and encouraging as I battled my own electrosensitivity while learning how to overcome it by writing this book!

And of course to James William Templeton, my partner, my love, my everything.

NOTES

Chapter 1:
The Body Electric

1. Robert O. Becker and Gary Selden, *The Body Electric: Electromagnetism and the Foundation of Life* (New York: Harper Paperbacks, 1998); Richard Gerber, *Vibrational Medicine, The #1 Handbook of Subtle-Energy Therapies* (Rochester, VT: Bear & Company, 2001).

2. William Tiller, *Psychoenergetic Science: A Second Copernican Scale Revolution* (Walnut Creek, CA: Pavior, 2007).

3. James Oschman, *Energy Medicine: The Scientific Basis* (Edinburgh: Churchill Livingstone, 2000).

4. National Research Council, *Severe Space Weather Events—Understanding Societal and Economic Impacts: A Workshop Report* (Washington, D.C.: National Academy of Sciences, 2009).

5. Stephen Sinatra, *Heart, Health & Nutrition* (June 2008): 7.

6. Curious about Astronomy?, http://curious.astro.cornell.edu/question.php?number=307.

7. Robert O. Becker, *Cross Currents: The Perils of Electropollution, The Promise of Electromedicine* (New York: Tarcher, 1990).

8. WordPress.com News, "Magnetic Fields," http://wordpress.com/tag/60-hz-magnetic-fields/feed (accessed September 24, 2007).

Chapter 2:
The Body Interrupted

1. S. Sadetzki et al, "Cellular Phone Use and Risk of Benign and Malignant Parotid Gland Tumors—A Nationwide Case Control Study," *American Journal of Epidemiology* 167, no. 4 (2008): 457–67.

2. U.S. Federal Radiofrequency Interagency Working Group (*Guidelines Statement,* June 1999).

3. Cindy Sage and David Carpenter, "Public Health Implications of Wireless Technologies," *Pathophysiology* 16, no. 2 (August 2009): 233–46.

4. J. O. Nriagu, "Saturnine Gout among Roman Aristocrats. Did Lead Poisoning Contribute to the Fall of the Empire?" *The New England Journal of Medicine* 308, no. 11 (1983): 660–63.

5. Nriagu, "Saturnine Gout."

6. Sara Shannon, *Technology's Curse: Diet for the Atomic Age* (Anchorage: Earthpulse Press, 1993).

7. Sadetzki et al, "Cellular Phone Use."

8. Samuel Milham, "Historical Evidence That Electrification Caused the 20th Century Epidemic of 'Diseases of Civilization,'" *Medical Hypotheses* 74, no. 2 (September 2009): 337–45.

9. B. Blake Levitt, *Electromagnetic Fields: A Consumer's Guide to the Issues and How to Protect Ourselves* (New York: Backinprint.com, 2007).

10. Levitt, *Electromagnetic Fields.*

11. Levitt, *Electromagnetic Fields*; J. L. Kirschvink and J. L. Gould, "Biogenic Magnetite as a Basis for Magnetic Field Detection in Animals," *Biosystems* 13, no. 3 (1981): 181–201; J. L. Kirschvink et al, "Magnetite in Human Tissues: A Mechanism for the Biological Effects of Weak ELF Magnetic Fields," *Bioelectromagnetics* 1 (1992): 101–13.

12. Levitt, *Electromagnetic Fields*.

13. Levitt, *Electromagnetic Fields*.

14. W. R. Adey, "Biological Effects of Electromagnetic Fields," *Journal of Cellular Biochemistry* 51, no. 4 (April 1993): 410–16; W. R. Adey, "Cell Membranes: The Electromagnetic Environment and Cancer Promotion," *Neurochemical Research* 13, no. 7 (July 1988): 671–77.

15. Adey, "Biological Effects"; Adey, "Cell Membranes."

16. Andrew Goldsworthy, "The Biological Effects of Weak Electromagnetic Fields" (unpublished research, 2007).

17. S. Bawin and W. R. Adey, "Sensitivity of Calcium Binding in Cerebral Tissue to Weak Environmental Electric Fields Oscillating at Low Frequency," *Proceedings of the National Academy of Science USA 1976*, June 73, no. 6: 1999–2003.

18. Goldsworthy, "The Biological Effects."

19. Goldsworthy, "The Biological Effects."

20. Goldsworthy, "The Biological Effects."

21. Goldsworthy, "The Biological Effects."

22. M. J. Abramson et al, "Mobile Telephone Use Is Associated with Changes in Cognitive Function in Young Adolescents," *Bioelectromagnetics* 30, no. 8 (2009): 678–86.

23. D. J. Panagopoulos et al, "Cell Death Induced by GSM 900-MHz and DCS 1800-MHz Mobile Telephony Radiation," *Mutation Research/Genetic Toxicology and Environmental Mutagenesis* 626, no. 1–2 (January 10, 2007): 69–78.

24. Levitt, *Electromagnetic Fields*.

25. H. Lai and N. P. Singh, "Magnetic-Field-Induced DNA Strand Breaks in Brain Cells of the Rat," *Environmental Health Perspectives* 112, no. 6 (May 2004): 687–94; H. Lai and N. P. Singh, "Acute Exposure to a 60 Hz Magnetic Field Increases DNA Strand Breaks in Rat Brain Cells," *Bioelectromagnetics* 18, no. 2 (1997): 156–65.

26. N. Wertheimer and E. Leeper, "Electrical Wiring Configurations and Childhood Cancer," *American Journal of Epidemiology* 109, no. 3 (March 1979): 273–84.

27. D. Savitz and W. T. Kaune, "Childhood Cancer in Relation to a Modified Residential Wire Code," *Environmental Health Perspectives* 101, no. 1 (April 22, 1993): 76–80.

28. Geoffrey Lean, "Electronic Smog," *The Independent* (May 7, 2006).

29. K. Mild et al, "Long-Term Use of Cellular Phones and Brain Tumours: Increased Risk Associated with Use for > or = 10 Years," *Occupational Environmental Medicine* 64, no. 9 (2007): 626–32.

30. Sadetzki et al, "Cellular Phone Use."

31. H. A. Divan, L. Kheifets, C. Obel, and J. Olson, "Prenatal and Postnatal Exposure to Cell Phone Use and Behavioral Problems in Children," *Epidemiology* 19, no. 4 (July 2008): 523–29.

32. Neil Cherry, "EMR Reduces Melatonin in Animals and People," *EMF Guru*, http://www.feb.se/EMFguru/Research/emf-emr/EMR-Reduces-Melatonin.htm; J. B. Burch et al, "Geomagnetic Activity and Human Melatonin Metabolite Excretion," *Neuroscience Letters* 438, no. 1 (June 12, 2008): 76–79; J. B. Burch et al, "Nocturnal Excretion of Urinary Melatonin Metabolite among Electric Utility Workers," *Scandinavian Journal of Work, Environment and Health* 24, no. 3 (1998): 183–89; R. J. Reiter, "Melatonin Suppression by Static and Extremely Low Frequency Electromagnetic Fields: Relationship to the Reported Increased Incidence of Cancer," *Review of Environmental Health* 10, no. 3–4 (July–December 1994): 171–86.

33. Barrie Lambert, "Radiation, Early Warning, Late Effect," Late Lessons from Early Warnings: The Precautionary Principle, 1896–2000, Environmental Issue Report, European Environment Agency, 2001, pp 17–27.

34. Lambert, "Radiation, Early Warning."

35. Lambert, "Radiation, Early Warning."

36. Levitt, *Electromagnetic Fields.*

37. BioInitiative, "BioInitiative Report: A Rationale for Biologically-Based Public Exposure Standard for Electromagnetic Fields (ELF and RF)," August 31, 2007, http://www.bioinitiative .org/report/index.htm.

38. E. D. Kirson et al, "Alternating Electric Fields Arrest Cell Proliferation in Animal Tumor Models and Human Brain Tumors," *Proceedings of the National Academy of Sciences USA* 104, no. 24 (June 12, 2007): 10152–57.

Chapter 3:
Lifting the Veil

1. S. C. Barber, P. J. Shaw, "Oxidative Stress in ALS: Key Role in Motor Neuron Injury and Therapeutic Target," *Free Radical Biology & Medicine* 48, no. 5 (2010): 629–41.

2. C. Y. Li and F. C. Sung, "Association between Occupational Exposure to Power Frequency Electromagnetic Fields and Amyotrophic Lateral Sclerosis: A Review." *American Journal of Indian Medicine* 43, no. 2 (February 2003): 212–20; *BioInitiative Report.*

3. C. Johansen, "Electromagnetic Fields and Health Effects—Epidemiologic Studies of Cancer, Diseases of the Central Nervous System and Arrhythmia-Related Heart Disease," *Scandinavian Journal of Work, Environment and Health* 30, no. 1 (2004): 1–30.

4. "Glutathione," *USANA Technical Bulletin,* January 2008.

5. "Chernobyl," *The Irish Times,* April 26, 2001; International Atomic Energy Agency, "Thyroid Cancer Effects in Children," August 2005 staff report; V. S. Kazakov et al, "Thyroid Cancer after Chernobyl," *Nature* 359, no. 6390 (September 2, 1992): 21; The American Thyroid Association, *Nuclear Radiation and the Thyroid,* 2005; Ellen J. Sullivan, "Chernobyl, 20 Years Later: ASCP Leader Heads Pathology Panel That Diagnoses Cancer Cases," *Medscape Pathology* (January 9, 2007).

6. A. Koyu et al, "Effects of 900 MHz Electromagnetic Field on TSH and Thyroid Hormones in Rats," *Toxicology Letters* 157, no. 3 (July 4, 2005): 257–62.

7. R. Seaberg et al, "Influence of Previous Radiation Exposure on Pathologic Features and Clinical Outcome in Patients with Thyroid Cancer," *Archives of Otolaryngology, Head and Neck Surgery* 135, no. 4 (2009): 355–59.

8. Cherry, "EMR Reduces Melatonin."

9. R. J. Reiter et al, "Light at Night, Chronodisruption, Melatonin Suppression, and Cancer Risk: A Review," *Critical Reviews Oncogenesis* 13, no. 4 (December 2007): 303–28.

10. Seaberg et al, "Influence of Previous Radiation."

11. Mast Sanity, "Germany: Study on Residents near Mast Shows Alarming Evidence of Harm to Health and Wellbeing," http://www.mastsanity.org/index.php?option=com_content&task=view&id=230&Itemid=136; H. Eger et al, "The Influence of Being Physically Near to a Cell Phone Transmission Mast on the Incidence of Cancer," *Umwelt·Medizin·Gesellschaft* 17, no. 4 (2004).

12. Levitt, *Electromagnetic Fields.*

13. D. Stopczyk et al, "Effect of Electromagnetic Field Produced by Mobile Phones on the Activity of Superoxide Dismutase (SOD-I) and the Level of Malonyldialdehyde (MDA)—In Vitro Study," *Medycyna Pracy* 53, no. 4 (2002): 311–14.

14. J. L. Phillips, W. D. Winters, and L. Rutledge, "In Vitro Exposure to Electromagnetic Fields: Changes in Tumour Cell Properties," *International Journal of Radiation Biology* 49, no. 3 (March 1986): 463–69.

15. *BioInitiative Report.*

16. Becker and Selden, *The Body Electric.*

17. Levitt, *Electromagnetic Fields.*

18. Levitt, *Electromagnetic Fields.*

Chapter 5:
Zap-Proof Your Home

1. Leora Boydal, "Compact Fluorescent Light Bulbs Draw Quality Complaints," *The New York Times*, March 27, 2009.

2. Lone Aussie, LLC, "Discovery of Earthing," http://whatisearthing.com/discovery_of_earthing (accessed March 4, 2010).

3. N. Wertheimer and E. Leeper, "Possible Effects of Electric Blankets and Heated Waterbeds on Fetal Development," *Bioelectromagnetics* 7, no. 1 (1986): 13–22.

4. C. Ober, R. Coghill, "Does Grounding the Human Body to Earth Reduce Chronic Inflammation and Chronic Pain?" (presentation, European Bioelectromagnetics Association, Budapest, Hungary, November 12, 2003).

5. M. Ghaly, D. Teplitz, "The Biologic Effects of Grounding the Human Body during Sleep as Measured by Cortisol Levels and Subjective Reporting of Sleep, Pain, and Stress," *Journal of Alternative and Complementary Medicine* 10, no. 5 (October 2004): 767–76.

6. G. Chevalier, K. Mori, and J. L. Oschman, "The Effect of Earthing (Grounding) on Human Physiology," *European Biology and Bioelectromagnetics* (January 31, 2006): 600–621.

7. C. Graham et al, "Heart Rate Variability and Physiological Arousal in Men Exposed to 60 Hz Magnetic Fields," *Bioelectromagnetics* 21, no. 6 (September 2000): 480–82.

8. Camilla Rees and Magda Havas, Public Health SOS. Boulder: Wide Angle Health, 2009.

Chapter 6:
Advanced Zap-Proofing

1. Ed Leeper, *Silencing the Fields: A Practical Guide to Reducing AC Magnetic Fields* (Boulder: Symmetry Books, 2001).

2. Leeper, *Silencing the Fields*; Riley, *Tracing EMFs in Building Wiring and Grounding* (Boulder: Magnetic Sciences, 2005).

3. Ian Sample, "How Does This Field of Lights Work?" *The Guardian*, February 26, 2004.

Chapter 7:
The Newest Zapper: Dirty Electricity

1. Magda Havas, Associate Professor of Environmental & Resource Studies at Trent University (Peterborough, Ontario, Canada).

2. Lloyd Morgan, "Blood Glucose Levels: A Study of Correlation Factors" (unpublished research, Brighton School, Brighton, WI, 2003.) This study looked at Stetzer's blood sugar.

3. "Hazards of Harmonics and Neutral Overloads," White Paper #26, American Power Conversion, 2003.

4. M. Havas, M. Illiatovitch, and C. Proctor, "Teacher and Student Response to the Removal of Dirty Electricity by the Graham/Stetzer Filter at Willow Wood School in Toronto, Canada" (presentation, Third International Workshop on the Biological Effects of Electromagnetic Fields, Kos, Greece, October 4–8, 2004): 311–17.

5. M. Havas and A. Olstad, "Power Quality Affects Teacher Well-Being and Student Behavior in Three Minnesota Schools," *The Science of the Total Environment* 401, no. 2–3 (September 1, 2008): 157–62.

6. M. Havas, "Electromagnetic Hypersensitivity: Biological Effects of Dirty Electricity with Emphasis on Diabetes and Multiple Sclerosis," *Electromagnetic Biology and Medicine* 25, no. 4 (2006): 259–68.

7. Riley, *Tracing EMFs*.

Chapter 8:
Zap-Proof Your Phone

1. C. Sage, O. Johansson, and S. A. Sage, "Personal Digital Assistant (PDA) Cell Phone Units Produce Elevated Extremely-Low Frequency Electromagnetic Field Emissions," *Bioelectromagnetics* 28, no. 5 (July 2007): 386–92.

2. George Carlo and Martin Schram, *Cell Phones: Invisible Hazards in the Wireless Age, An Insider's Alarming Discoveries about Cancer and Genetic Damage* (Laguna Beach: Basic Books, 2002).

3. Carlo and Schram, *Cell Phones.*

4. A. Huss et al, "Source of Funding and Results of Studies of Health Effects of Mobile Phone Use: Systematic Review of Experimental Studies," *Environmental Health Perspectives* 115, no. 1 (January 2007): 1–4.

5. University of Pittsburgh Department of Epidemiology, "Caution on Cell Phone Use," http://www.epidemiology.pitt.edu/news/News_Display.asp?link=357.

6. L. Hardell, M. Carlberg, and K. Hansson Mild, "Pooled Analysis of Two Case-Control Studies on the Use of Cellular and Cordless Telephones and the Risk of Benign Brain Tumours Diagnosed during 1997–2003," *International Journal of Oncology* 28, no. 2 (February 2006): 509–18.

7. James Niccolai, "Cell Phones Don't Cause Tumors, Study Finds," *PC World,* August 31, 2005, http://www.pcworld.com/article/122382/cell_phones_dont_cause_tumors_study_finds .html.

8. L. Hardell and M. Carlberg, "Mobile Phones, Cordless Phones and the Risk for Brain Tumours," *International Journal of Oncology* 35, no. 1 (July 2009): 5–17.

9. Mild et al, "Long-Term Use of Cellular Phones."

10. H. Lai and N. P. Singh, "Melatonin and N-tert-butyl-alpha-phenylnitrone Block 60-Hz Magnetic Field-Induced DNA Single and Double Strand Breaks in Rat Brain Cells," *Journal of Pineal Research* 22, no. 3 (April 1997): 152–62.

11. H. Lai and N. P. Singh, "Magnetic-Field-Induced DNA Strand Breaks in Brain Cells of the Rat," *Environmental Health Perspectives* 112, no. 6 (May 2004): 687–94.

12. H. Lai. "Evidence for Genotoxic Effects (RFR and ELF Genotoxicity)," *BioInitiative Report,* July 2007.

13. Lai, "Evidence for Genotoxic Effects."

14. Sage et al, *BioInitiative Report,* July 2007.

15. R. Wolf and D. Wolf, "Increased Incidence of Cancer Near a Cell-Phone Transmitter Station," *International Journal of Cancer Prevention* 1, no. 2 (April 2004).

16. M. Blettner et al, "Mobile Phone Base Stations and Adverse Health Effects: Phase 1 of a Population-Based, Cross-Sectional Study in Germany" *Occupational and Environmental Medicine* 66, no. 2 (February 2009): 118–23.

17. Christopher Ketcham, "Warning: Your Cell Phone May Be Hazardous to Your Health," *GQ,* February 2010.

18. A. Agarwal et al, "Effect of Cell Phone Usage on Semen Analysis in Men Attending Infertility Clinic: An Observational Study," *Fertility and Sterility* 89, no. 1 (January 2008): 124–28.

19. M. Hours et al, "Cell Phones and Risk of Brain and Acoustic Nerve Tumours: The French INTERPHONE Case—Control Study," *Revue d'Épidémiologie et de Santé Publique* 55, no. 5 (October 2007): 321–32.

20. C. Sage, O. Johansson, and S. A. Sage, "Personal Digital Assistant (PDA) Cell Phone Units Produce Elevated Extremely-Low Frequency Electromagnetic Field Emissions," *Bioelectromagnetics* 28, no. 5 (July 2007): 386–92.

21. A. Agarwal, "Cell Phones: Modern Men's Nemesis," Reproductive BioMedicine Online, November 3, 2008.

22. E. Cardis et al, "The Interphone Study: Design, Epidemiological Methods, and Description of the Study Population," *European Journal of Epidemiology* 22, no. 9 (2007): 647–64.

23. Cardis et al, "The Interphone Study."

24. Microwave News, "Interphone Project: The Cracks Begin to Show," June 19, 2008, http://www.microwavenews.com/interphonecracks.html; Doreen Carvajal, "Rift Delays Official Release of Study on Safety of Cellphones," *The New York Times,* June 29, 2008.

25. "Interphone Points to Long-Term Brain Tumor Risks Interpretation Under Dispute" (*Microwave News,* May 17, 2010 (update May 18); http://www.microwavenews.com/Interphone.Main.html.

Chapter 9:
Zap-Proof Your Kids

1. Clean Air Task Force, "Children at Risk: How Air Pollution from Power Plants Threatens the Health of America's Children," May 2002, http://www.catf.us/publications/view/14.

2. A. D. Tinniswood, C. M. Furse, and O. P. Gandhi. "Computations of SAR Distributions for Two Anatomically-Based Models of the Human Head Using CAD Files of Commercial Telephones and the Parallelized FDTD Code," *IEEE Transactions on Antennas and Propagation* 46 (June 1998): 829–33; Jianqing Wang and O. Fujiwara, "Comparison and Evaluation of Electromagnetic Absorption Characteristics in Realistic Human Head Models of Adult and Children for 900-MHz Mobile Telephones," *IEEE Transactions on Microwave Theory and Techniques* 51, no. 3 (March 2003): 966–71.

3. Maisch, "Children and Mobile Phones," www.avaate.org/IMG/pdf/030709_Maisch_Children_Mobile-Phones.pdf ("What Cell Phones Can Do to Youngster's Brain in 2 Minutes," *UK Sunday Mirror,* April 1, 2004.

4. L. G. Salford et al, "Non-Thermal Effects of EMF upon the Mammalian Brain" (presentation, The Precautionary EMF Approach: Rationale, Legislation and Implementation international conference, Benevento, Italy, February 2007).

5. "Sensitivity of Children to EMF Exposure" (workshop proceedings, World Health Organization, Ankara, Turkey, June 9–10, 2004).

6. "Sensitivity of Children to EMF Exposure."

7. Wireless Protection, "Body Voltage," www.wireless-protection.org/electrosmog_02.html.

8. Divan et al, "Prenatal and Postnatal Exposure."

9. D. K. Li et al, "A Population-Based Prospective Cohort Study of Personal Exposure to Magnetic Fields during Pregnancy and the Risk of Miscarriage," *Epidemiology* 13, no. 1 (January 2002): 9–20.

10. M. P. de la Puente and A. Balmori, "Addiction to Cell Phones: Are There Neurophysiological Mechanisms Involved?" *Proyecto* 61 (March 2007): 8–12.

11. De la Puente and Balmori, "Addiction to Cell Phones."

12. H. Lai et al, "Intraseptal Microinjection of Beta-Funaltrexamine Blocked a Microwave-Induced Decrease of Hippocampal Cholinergic Activity in the Rat," *Pharmacology, Biochemistry, and Behavior* 52, no. 3 (1996): 613–16.

13. D. O. Carpenter and C. Sage, "Setting Prudent Public Health Policy for Electromagnetic Field Exposures," *Reviews on Environmental Health* 23, no. 2 (April–June 2008): 91–117.

14. G. Gandhi, "Genetic Damage in Mobile Phone Users: Some Preliminary Findings." *Indian Journal of Human Genetics,* 2005, 11: 99–104.

15. N. K. Panda et al, "Audiologic Disturbances in Long-Term Mobile Phone Users" (presentation, American Academy of Otolaryngology-Head and Neck Surgery Foundation annual meeting & OTO EXPO, Washington, D.C., September 2007).

16. Andrew Goldsworthy, "Electromagnetic Fields and Health: Executive Report." EM Radiation Research Trust, 2009.

17. Richard Shim, "Parents Sue School District for Wi-Fi Use," Cnet News, October 9, 2003.

18. Joanna Bale, "Health Fears Lead Schools to Dismantle Wireless Networks," *Times Online*, November 20, 2006.

19. Andrea L. Foster, "College Librarian Quits After Citing Health Concerns Related to Wireless Network," *The Chronicle of Higher Education*, February 2, 2007.

20. Joe Moszczynski, "Fredon Closes Lone School over Power Line Concerns," *The Star Ledger*, September 9, 2009.

21. Moszczynski, "Fredon Closes."

22. Robert Sylers, "EMF and Childhood Leukemia," *Electrical Construction & Maintenance*, September 1, 2006.

23. Sylers, "EMF and Childhood Leukemia."

24. R. Kavet et al, "Association of Residential Magnetic Fields with Contact Voltage," *Bioelectromagnetics* 25, no. 7 (October 2004): 530–36; R. Kavet and L. E. Zaffanella, "Contact Voltage Measured in Residences: Implications to the Association between Magnetic Fields and Childhood Leukemia," *Bioelectromagnetics* 23, no. 6 (September 2002): 464–74; R. Kavet et al, "The Possible Role of Contact Current in Cancer Risk Associated with Residential Magnetic Fields," *Bioelectromagnetics* 21, no. 7 (October 2000): 538–53.

25. C. Rice, "Prevalence of Autism Spectrum Disorders—Autism and Developmental Disabilities Monitoring Network, United States, 2006," *Morbidity and Mortality Weekly Report* 18 (December 2009).

26. Goldsworthy, "The Biological Effects."

27. Levitt, *Electromagnetic Fields*.

28. I. M. Thornton, "Out of Time: A Possible Link between Mirror Neurons, Autism and Electromagnetic Radiation," *Medical Hypotheses* 67, no. 2 (2006): 378–82.

29. Thornton, "Out of Time."

30. T. J. Mariea and G. L. Carlo, "Wireless Radiation in the Etiology and Treatment of Autism: Clinical Observations and Mechanisms," *Journal of the Australasian College of Nutritional and Environmental Medicine* 26, no. 2 (August 2007): 3–7.

Chapter 10:
Zap-Proof Your Work Environment

1. U.S. Department of Health and Human Services, "NIOSH Fact Sheet: EMFs in the Workplace," DHHS (NIOSH) Publication No. 96–129, http://www.cdc.gov/niosh/emf2.html.

2. L. E. Charles et al, "Electromagnetic Fields, Polychlorinated Biphenyls, and Prostate Cancer Mortality in Electric Utility Workers," *American Journal of Epidemiology* 157, no. 8 (2003): 683–91.

3. E. van Wijngaarden et al, "Exposure to Electromagnetic Fields and Suicide among Electric Utility Workers: A Nested Case-Control Study," *Western Journal of Medicine* 173, no. 2 (August 2000): 94–100.

4. S. Perry, L. Pearl, R. Binns, "Power Frequency Magnetic Field; Depressive Illness and Myocardial Infarction," *Public Health* 103, no. 3 (1998): 177–80.

5. P. Demers et al, "Occupational Exposure to Electromagnetic Fields and Breast Cancer in Men," *American Journal of Epidemiology* 134, no. 4 (1991): 340–347.

6. T. Tynes, A. Andersen, and F. Langmark, "Incidence of Cancer in Norwegian Workers Potentially Exposed to Electromagnetic Fields," *American Journal of Epidemiology* 136, no. 1 (1992): 81–88.

7. National Cancer Institute, "General Information about Male Breast Cancer," www.cancer.gov/cancertopics/pdq/treatment/malebreast/Patient.

8. Tara Parker-Pope, "Campus Building Blamed for Cancer Cluster," *The New York Times*, February 24, 2009.

9. Parker-Pope, "Campus Building Blamed."

10. S. Milham and L. Morgan, "Teachers' Cancer Cluster at La Quinta Middle School," unpublished report, April 2007.

11. Havas, Illiatovitch, and Proctor, "Teacher and Student Response"; M. Havas, "Dirty Electricity: An Invisible Pollutant in Schools," *Education Forum Magazine, OSSTF* 32, no. 3 (2006).

12. A. Wilkins et al, "Fluorescent Lighting, Headaches and Eyestrain," *Lighting Research and Technology* 21, no. 1 (1989): 11–18.

13. Agarwal, "Effect of Cell Phone."

14. R. A. Buckin et al, "Stray Voltage in Dairies," University of Florida IAFS Extension, 2009.

Chapter 11:
Other Zappers in Your Life

1. D. A. Schauer and O. W. Lintton, "National Council on Radiation Protection and Measurement Report Shows Substantial Medical Exposure Increase," *Radiology* 253 (2009): 293–96.

2. D. J. Brenner and E. J. Hall, "Computed Tomography—An Increasing Source of Radiation Exposure," *The New England Journal of Medicine* 357, no. 22 (November 2007): 2277–84; R. Fazel et al, "Exposure to Low-Dose Ionizing Radiation from Medical Imaging Procedures," *The New England Journal of Medicine* 361, no. 9 (August 27, 2009): 849–57; R. Smith-Bindman et al, "Radiation Dose Associated with Common Computed Tomography Examinations and the Associated Lifetime Attributable Risk of Cancer," *Archives of Internal Medicine* 169, no. 22 (December 14, 2009): 2078–86.

3. E. S. Amis et al, "American College of Radiology White Paper on Radiation Dose in Medicine." *Journal of the American College of Radiology,* 2007; 4: 272–84.

4. R. Fazel et al, "Exposure to Low-Dose Ionizing Radiation."

5. Shankar Vedantem, "Doctors Reap Benefits by Doing Own Tests," *The Washington Post,* July 31, 2009.

6. R. J. Dachs, M. A. Graber, and A. Darby-Stewart, "Cancer Risks Associated with CT Scanning," *American Family Physician* 81, no. 2 (January 15, 2010); D. L. Preston et al, "Studies of Mortality of Atomic Bomb Survivors. Report 13: Solid Cancer and Noncancer Disease Mortality: 1950–1997," *Radiation Research* 160, no. 4 (October 2003): 381–407.

7. U.S. Environmental Protection Agency, "Radiation Protection," http://www.epa.gov/rpdweb00/.

8. Public Citizen, "The Case Against Nuclear Power," http://www.citizen.org/cmep/energy_enviro_nuclear/nuclear_power_plants/.

9. Jay M. Gould, *The Enemy Within: The High Cost of Living Near Nuclear Reactors* (New York: Four Walls Eight Windows, 1996).

10. J. Mangano and J. D. Sherman, "Childhood Leukaemia near Nuclear Installations," *European Journal of Cancer Care* 17, no. 4 (July 2008): 416–18.

11. M. Donohoe, "Unnecessary Testing in Obstetrics, Gynecology, and General Medicine: Causes and Consequences of the Unwarranted Use of Costly and Unscientific (Yet Profitable) Screening Modalities," *Medscape Today,* April 30, 2007.

12. American College of Radiology, http://www.acr.org.

13. American College of Radiology, http://www.acr.org.

14. Preston, "Studies of Mortality."

15. C. Junghans and A. D. Timmins, "Risk Assessment after Acute Coronary Syndrome," *British Medical Journal* 333, no. 7578 (November 25, 2006): 1079–80.

16. U.S. Preventive Services Task Force, "Screening for Coronary Heart Disease," Agency for Healthcare Research and Quality, February 2004.

17. R. P. Jensh and R. L. Brent, "Intrauterine Effects of Ultrasound: Animal Studies," *Teratology* 59, no. 4 (1997): 240–51.

18. U.S. Environmental Protection Agency, "Radon," http://www.epa.gov/radon.

19. U.S. Environmental Protection Agency, "Radon."

Chapter 12:
Zap-Proof Superfoods and Seasonings

1. Nutrient Data Laboratory, Beltsville Human Nutrition Research Center, Agricultural Research Service, and U.S. Department of Agriculture, "Oxygen Radical Absorbance Capacity (ORAC) of Selected Foods—2007," November 2007, http://www.ars.usda.gov/sp2userfiles/place/12354500/data/orac/orac07.pdf.

2. C. Nencini, G. Giorgi, and L. Micheli, "Protective Effect of Silymarin on Oxidative Stress in Rat Brain." *Phytomedicine* 14, no. 2–3 (February 2007): 129–35.

3. R. Ferracane et al, "Effects of Different Cooking Methods on Antioxidant Profile, Antioxidant Capacity, and Physical Characteristics of Artichoke," *Journal of Agricultural Food Chemistry,* September 28, 2008.

4. C. S. Bediz et al, "Zinc Supplementation Ameliorates Electromagnetic Field-Induced Lipid Peroxidation in the Rat Brain," *The Tohoku Journal of Experimental Medicine* 208, no. 2 (February 2006): 133–40.

5. S. Kirkham et al, "The Potential of Cinnamon to Reduce Blood Glucose Levels in Patients with Type 2 Diabetes and Insulin Resistance," *Diabetes, Obesity & Metabolism* 11, no. 12 (December 2009): 1100–13.

6. J. Joseph and B. Shukitt-Hale, unpublished research.

7. F. Ozguner et al, "Mobile Phone-Induced Myocardial Oxidative Stress: Protection by a Novel Antioxidant Agent Caffeic Acid Phenethyl Ester," *Toxicology and Industrial Health* 21, no. 9 (October 2005): 223–30.

8. Ozguner et al, "Mobile Phone-Induced," 223–30.

9. K. Aruna and V. M. Sivaramakrishnan, "Anticarcinogenic Effects of the Essential Oils from Cumin, Poppy and Basil," *Phytotherapy Research* 10, no. 7 (1998): 577–80.

10. R. S. Farag and K. H. el-Khawas, "Influence of Gamma-Irradiation and Microwaves on the Antioxidant Property of Some Essential Oils," *International Journal of Food Sciences and Nutrition* 49, no. 2 (March 1998): 109–15.

11. R. Thomas, "Changing Genes: Garlic Shown to Inhibit DNA Damaging Chemical in Breast Cancer" (presentation, Frontiers in Cancer Prevention Research, American Association for Cancer Research, Baltimore, MD, 2005).

12. J. M. Leheska et al, "Effects of Conventional and Grass-Feeding Systems on the Nutrient Composition of Beef," *Journal of Animal Science* 86 (2008): 3575–85.

13. N. J. Dubost et al, "Identification and Quantification of Ergothioneine in Cultivated Mushrooms by Liquid Chromatography-Mass Spectroscopy," *International Journal for Medicinal Mushrooms* 8 (2006): 215–22.

14. "Oxygen Radical Absorbance Capacity (ORAC) of Selected Foods—2007." Prepared by Nutrient Data Laboratory, Beltsville Human Nutrition Research Center (BHNRC), Agricultural Research Service (ARS), U.S. Department of Agriculture (USDA) in collaboration with Arkansas Children's Nutrition Center, ARS, USDA, Little Rock, AR, November 2007.

15. M. D. Kontogianni et al, "The Impact of Olive Oil Consumption Pattern on the Risk of Acute Coronary Syndromes: The CARDIO2000 Case-Control Study," *Clinical Cardiology* 30, no. 3 (March 2007): 125–29.

16. G. K. Beauchamp et al, "Phytochemistry: Ibuprofen-Like Activity in Extra-Virgin Olive Oil," *Nature* 437, no. 7055 (September 1, 2005): 45–46.

17. F. Caconio et al, "Influence of the Exposure to Light on Extra Virgin Olive Oil Quality on Storage," European Food Research and Technology, 2005, 221(1–2): 92–98.

18. N. P. Seeram et al, "Comparison of Antioxidant Potency of Commonly Consumed Polyphenol-Rich Beverages in the United States," *Journal of Agricultural and Food Chemistry* 56, no. 4 (February 27, 2008): 1415–22.

19. M. Aviram et al, "Pomegranate Juice Consumption Reduces Oxidative Stress, Atherogenic Modifications to LDL, and Platelet Aggregation: Studies in Humans and in Atherosclerotic Apolipoprotein E-Deficient Mice," *The American Journal of Clinical Nutrition* 71 (May 2000): 1062–76.

20. A. Szuchman et al, "Characterization of Oxidative Stress in Blood from Diabetic vs. Hypercholesterolaemic Patients, Using a Novel Synthesized Marker," *Biomarkers* 13, no. 1 (February 2008): 119–31.

21. K. Nakatani et al, "Inhibition of Cyclooxygenase and Prostaglandin E2 Synthesis by Gamma-Mangostin, a Xanthone Derivative in Mangosteen, in C6 Rat Glioma Cells," *Biochemical Pharmacology* 63, no. 1 (January 2002): 73–79.

22. College of Tropical Agriculture and Human Resources, "The Noni Website," http://www.ctahr.hawaii.edu/noni/.

23. A. Jindal et al, "Radioprotective Potential of *Rosemarinus officinalis* against Lethal Effects of Gamma Radiation: A Preliminary Study," *Journal of Environmental Pathology, Toxicology and Oncology* 25, no. 4 (2006): 633–34.

24. D. Soyal et al, "Modulation of Radiation-Induced Biochemical Alterations in Mice by Rosemary (*Rosemarinus officinalis*) Extract," *Phytomedicine* 14, no. 10 (October 2007): 701–5.

25. www.ncbi.nlm.nih.gov/pmc/articles/PMC1930474/pdf/canmedaj01601-0045.pdf.

26. S. Burkhardt et al, "Detection and Quantification of the Antioxidant Melatonin in Montmorency and Balaton Tart Cherries (*Prunus cerasus*)," *Journal of Agricultural and Food Chemistry* 49, no. 10 (2001): 4898–4902.

27. K. S. Kuehl et al, "Efficacy of Tart Cherry Juice in Reducing Muscle Pain During Running: A Randomized Controlled Trial." *Journal of the International Society of Sports Medicine* (May 2010): 7–17.

28. M. Nagabhushan and S. V. Bhide, "Curcumin as an Inhibitor of Cancer," *Journal of the American College of Nutrition* 11, no. 2 (April 1992): 192–98.

29. B. B. Aggarwal, A. Kumar, and A. C. Bharti, "Anticancer Potential of Curcumin: Preclinical and Clinical Studies," *Anticancer Research* 23, no. 1A (January–February 2003): 363–98.

30. G. P. Lim et al, "The Curry Spice Curcumin Reduces Oxidative Damage and Amyloid Pathology in an Alzheimer Transgenic Mouse," *The Journal of Neuroscience* 21, no. 21 (November 1, 2001): 8370–77.

31. S. Y. Park and D. S. Kim, "Discovery of Natural Products from *Curcuma longa* That Protects Cells from Beta-Amyloid Insult: A Drug Discovery Effort Against Alzheimer's Disease," *Journal of Natural Products* 65, no. 9 (September 2002): 1227–31.

32. T. O. Khor et al, "Combined Inhibitory Effects of Curcumin and Phenethly Isothiocyanate on the Growth of Human PC-3 Prostate Xenografts in Immunodeficient Mice," *Cancer Research* 66, no. 2 (January 15, 2006): 613–21.

33. M. Johnson et al, "Omega-3/Omega-6 Fatty Acids for Attention Deficit Hyperactivity Disorder: A Randomized Placebo-Controlled Trial in Children and Adolescents," *Journal of Attention Disorders* 12, no. 5 (March 2009): 394–401.

34. A. Goldsworthy, "The Biological Effects of Weak Electromagnetic Fields," 2007, http://www.radiationresearch.org/pdfs/goldsworthy_bio_weak_em_07.pdf.

Chapter 13:
Zap-Proof Minerals and Supplements

1. J. K. Gammack and J. M. Burke, "Natural Light Exposure Improves Subjective Sleep Quality in Nursing Home Residents," *Journal of the American Medical Directors Association* 10, no. 6 (July 2009): 440–41.

2. T. Harada, "Effects of Evening Light Conditions on Salivary Melatonin of Japanese Junior High School Students," *Journal of Circadian Rhythms* 2, no. 4 (2004).

3. V. Klinkenborg. "Our Vanishing Night," *National Geographic* (November 2008).

4. L. Kayumov et al, "Blocking Low-Wavelength Light Prevents Nocturnal Melatonin Suppression with No Adverse Effect on Performance During Simulated Shift Work," *Journal of Clinical Endocrinology & Metabolism* 90, no. 5 (2005): 2755–61.

5. M. Havas, "Electromagnetic Hypersensitivity"; M. Havas, "Dirty Electricity, Diabetes and Multiple Sclerosis" (presentation, Centre for Health Studies Research, Trent University, Peterborough, ON, January 25, 2006).

6. S. Amara et al, "Zinc Supplementation Ameliorates Static Magnetic Field-Induced Oxidative Stress in Rat Tissues," *Environmental Toxicology and Pharmacology* 23, no. 2 (March 2007): 193–97.

7. N. Buscemi et al, "Melatonin for Treatment of Sleep Disorders," Evidence Report/Technology Assessment: Number 108, Agency for Healthcare Research and Quality (November 2004).

8. D. P. Hayes, "The Protection Afforded by Vitamin D Against Low Radiation Damage," *International Journal of Low Radiation* 5, no. 4 (2008): 368–94.

9. A. Valenzuela et al, "Selectivity of Silymarin on the Increase of the Glutathione Content in Different Tissues of the Rat," *Planta Medica* 55, no. 5 (October 1989): 420–22.

10. F. Kadrnka, "Results of a Multicenter Orgotein Study in Radiation Induced and Interstitial Cystitis," *European Journal of Rheumatology and Inflammation* 4, no. 2 (1981): 237–43.

11. Ozguner et al, "Mobile Phone-Induced."

12. A. K. Johansson et al, "Sea Buckthorn Berry Oil Inhibits Platelet Aggregation," *The Journal of Nutritional Biochemistry* 11, no. 10 (October 2000): 491–95.

Epilogue

1. "Saturated Mobile Networks—Breaking Up, Will the Rapid Growth in Data Traffic Overwhelm Wireless Networks?" *The Economist* (February 11, 2010); http://www.economist.com/node/15498399?story_id-15498399 (accessed June 21, 2010).

Zapped Resources, Support, and Solutions

1. J. R. Cram, "Effects of Two Flower Essences on High Intensity Environmental Stimulation and EMF: A Matter of Head and Chest," *Subtle Energies and Energy Medicine Journal* 12, no. 3 (2001).